Over the Rainbow

Alex James is a writer, farmer, and the bass player with Blur. He lives in Oxfordshire with his wife and five children.

Over the Rainbow

Tales from an Unexpected Year

ALEX JAMES

PARTICULAR BOOKS
an imprint of
PENGUIN BOOKS

PARTICULAR BOOKS

UK | USA | Canada | Ireland | Australia
India | New Zealand | South Africa

Particular Books is part of the Penguin Random House group of companies
whose addresses can be found at global.penguinrandomhouse.com.

Penguin Random House UK,
One Embassy Gardens, 8 Viaduct Gardens, London SW11 7BW

penguin.co.uk

Penguin
Random House
UK

First published in Particular Books 2024

001

Text copyright © Alex James, 2024

The moral right of the author has been asserted

Penguin Random House values and supports copyright.
Copyright fuels creativity, encourages diverse voices, promotes freedom
of expression and supports a vibrant culture. Thank you for purchasing
an authorized edition of this book and for respecting intellectual property
laws by not reproducing, scanning or distributing any part of it by any
means without permission. You are supporting authors and enabling
Penguin Random House to continue to publish books for everyone.
No part of this book may be used or reproduced in any manner for the
purpose of training artificial intelligence technologies or systems. In accordance
with Article 4(3) of the DSM Directive 2019/790, Penguin Random House
expressly reserves this work from the text and data mining exception.

Set in 11.75/13.75pt Dante MT Std
Typeset by Jouve (UK), Milton Keynes
Printed and bound in Great Britain by Clays Ltd, Elcograf S.p.A.

The authorized representative in the EEA is Penguin Random House Ireland,
Morrison Chambers, 32 Nassau Street, Dublin D02 YH68

A CIP catalogue record for this book is available from the British Library

ISBN: 978-0-241-71555-0

Penguin Random House is committed to a sustainable future
for our business, our readers and our planet. This book is made from
Forest Stewardship Council® certified paper.

MIX
Paper | Supporting
responsible forestry
FSC® C018179

For Claire, obviously. With my love.

Contents

1. Last Christmas — 1
2. Britpop Trousers — 15
3. In the Studio — 29
4. Half-Term — 41
5. Easter, Full Beluga — 55
6. Ten Hours in New York — 75
7. Showtime — 83
8. Going Large — 99
9. Wembley — 109
10. High Summer — 121
11. All Back to Mine — 135
12. Totally Cherry — 149
13. Further Afield — 161
14. Finale — 167
15. Next Year — 181

Acknowledgements — 185

1.

Last Christmas

December

Day to day, week to week, I'm consumed with the family farm. My wife and I bought it on our honeymoon twenty years ago, and it's our home and it's our business.

It was then such a rewilded heap that all the local billionaires and grand dynastic families of the parish – who, as a matter of course, seek to extend their preserves and will scrap, tooth and nail, over any farms in the neighbourhood that come on to the market – uniquely, universally turned their noses up at this one: a giddy sprawl of rambling, crumbling buildings, some that had once, long ago, been beautiful and more that would never be. It was more like buying a village than a house – an abandoned village at that, or perhaps a tiny, bankrupt principality. (Actually, it's about half the size of Monaco, which is titchy as farms go, although, unlike Monaco, no one rich had ever lived here.)

But it felt spiritually sound. Sacred, even. From the moment we first rolled up the front track through scattering rabbits, we both wanted to stay.

And the more time I spend here, with Claire, making it whirr, the less I want to leave. It constantly surprises me with its bounties and beauties, and it is, perhaps, most beautiful of all in darkest winter, under blankets of snow, carpets of frost, roofless and brilliant blue skies, bonfires blazing. I can go for weeks, months at a time, consumed by this place, with work and with family, with the people who make the farm tick.

But then, each year, Christmas comes, and we officially shut up shop. The office closes the day before Christmas Eve, and everyone has to take a break until the day after New Year's Day. No meetings, no scheming, no digging, no building, and no travel beyond the parish boundary.

'Never tell anyone we're closed for the holiday. Not ever,' I tell them on the last day in the office: Sian, the bookkeeper and first Lord of the Farm Treasury, Charlotte, the farm manager, and Georgie, who is good at absolutely everything but wants to leave and work somewhere else. 'Say we're travelling, say we're away, say anything but "we're on holiday" – that just annoys everyone who isn't.'

Since we've lived here, bang in the bullseye of England's green and pleasant land, I've realized that summer is just about the stupidest time to take a holiday, but then there's never a good time to leave a farm. Except, perhaps, the darkest, coldest couple of weeks.

There are always animals to feed, and the odd email to send, but everything else just stops at Christmas time. Probably always has, in fact, way before baby Jesus arrived – but nowadays December's feasting and merrymaking is so well established that it's impossible to get anything done for about a fortnight, even if you try. Not round here. You can't fight it.

Last Christmas

And so it was written that all those who live hereabouts would all be merrily ensconced, *en château, en famille et en fête*.

~

I'd been working hard, and had a huge mountain to climb in January. I was much looking forward to some well-earned rest and relaxation. Except, this year, Christmas Eve fell on a Saturday, and the entire week leading up to it was starting to look like one long Christmas party, a slow-motion steeplechase all over the gardens. Invitations had been trickling in for weeks.

The festivities kicked off with a birthday party on the shortest day of the year. Giddy with the anticipation of a week or two's repose and revelling, I got stuck right into the Armagnac during the canapés on arrival. (I'm normally a cider brandy man, but I switch to Armagnac at Christmas. Try it – but maybe not with the canapés.)

It was a small gathering: a sit-down dinner in a be-baubled orangery – although we didn't sit down to dinner until quite late, and after the toasts and the speeches, and the petits fours and the cheese and most of the Armagnac, Jamie Cullum played. With his whole band. There were, I think, almost as many people in the band as there were having dinner.

They were absolutely brilliant. I was dancing on the table before the first song had finished.

And then Jamie passed me the microphone and I sang 'Uptown Funk'.

Apparently.

Although I only have the faintest dream-like recollection of that part of the evening, I can tell you for certain that the mic would have been a Shure SM 58 unidirectional (cardioid) dynamic model.

It gave me pause to consider that something that I thought might never happen again, something very precious that I had only the faintest, dream-like recollection of, was about to be thrust upon me in a similar fashion.

But there was no time to dwell because the next day we were off – after a spot of family magnet fishing in the River Evenlode – to a candlelit carol service in a contemporized castle, complete with its own immaculately preserved ancient church.

After hitting the descant harmonies in the last verse of 'Hark the Herald' with my children in the back row, I found myself in the stateroom with a fathom of cider brandy, smoking a cigar about a foot long and wondering where to knock off the huge faggot of ash that was about to plop.

The stateroom was chock full of paintings by L.S. Lowry and *objets* and glacé fruits and Etonians. There was a gleaming crystal bowl on the ottoman I was relaxing on but I wasn't sure what it was.

I spotted the *châtelaine* and waved my brandy at the bowl, only spilling a little bit. 'Is this an ashtray, babe?'

'Everything in here's an ashtray, love.' She winked, waving back a glowing Marlboro Gold, white tip.

It was very relaxed.

The choir had decided to stay, *en masse*, and began singing round the ottoman, still in their robes and regalia. So I got to do the descant harmonies again. It ended up being quite a late one.

~

Due to residual cider brandy in the system, I had to take the train into Oxford the next morning in order to stock up on Christmas goodies. Fenton, a ballerina, was arriving shortly, as well as Auntie Maureen, so we needed to take on supplies.

We're a big family. Me, Claire, five kids and two grannies. It's a bit like doing Christmas every day, round here. We need two turkeys just for Sunday lunch. We do all our shopping wholesale. Not that we have to do it very often, because we live in a food machine.

A lot of the Christmas haul was already in. We keep pigs, so we had enough hams and bacons. There was a carrier bag full of summer truffles in the freezer, left over from some wild miscalculations at Feastival, the food and music festival we run on the farm each year. Cheese was covered, too, and in the market garden there was celery and celeriac, Jerusalem artichokes, black radishes, leeks and brassicas all up the yin yang. Plus, we'd already made a shedload of sugar plums for Fenton the ballerina.

But what the kids really wanted was Pot Noodles and Frazzles and Coca-Cola and stroopwafels.

Beatrix, our youngest, had been carefully planning the spree for some time. She's the quartermaster, in charge of the stores. She led me around the aisles at Booker Wholesale ticking items off her long, loving meticulously compiled list: 25-kilo sack of sugar for baking, sack of plain flour for baking, sack of oo flour for the boys (for pizza), self-raising flour (for her sister), cases of various ice creams, jar of Parma Violets, box of KitKats, 'and Baileys *for the grannies,*' she sang.

Some people spend their time and money on boats; some people spend their money on cars. I've always spent my money on food. Even when I didn't have any.

That's what money is for. Food – and drivers. A driver met us and took us home, fully loaded. Beatrix repaired to the present-wrapping department: the old pig shed. I went to the office to check my emails, but it was full of dancing teenagers, mostly my own. They were going loud and hard.

Fenton arrived then and was called upon immediately to judge the 'Rasputin' competition and then she started dancing and then Claire joined in and everyone started trying to do the splits and I had to go to bed because there was a shoot lunch the next day.

~

A shoot is a big-budget, big-footprint production, and it needs to be approached with caution. There are beaters and there are dogs and there's always all these lovely sausage rolls. There's inevitably a fair sprinkling of toffs and grandees among the guns, and a gratuitous famous person, probably James Blunt or Roger Waters, if you're really unlucky. But not me, because I don't really like it any more.

I do love stomping around in wellies in the fresh air, though. Even in sluicing rain and sideways snow, it's exhilarating. I like the chat and the bonfires. And I love the strict dress code – I always wear a brightly coloured tracksuit. And if there's a good reason to go along beyond any of those, it's that the birds are delicious.

The Christmas shoot was within walking distance, but a shooting party normally involves driving the night before for hours to a vast demesne somewhere you've never ever been or even ever heard of, where there is a maid with a teetering salver of liquid temptations waiting just inside the strap-hinged, double oak doors. Then there's cocktails and then there's white with the fish and red with the meat and a vase of port flying around in circles and hot brandy with the cigars and then they all want a sing-song round the Steinway and you've got off lightly if you're in bed by two.

I'd arranged for Geronimo, our eldest, to come and fire the gun for me. The previous day had technically been a designated *spa day*, but there were still the remains of the day-prior-to-that's hangover and a cumulative lack of sleep to contend with.

These were well met by an exceptionally pretty day. Air so cold you had to sip at it gently. A landscape frozen still, bottomless silences cracked by footsteps on frost, ricocheting laughter and Geronimo and I singing Beatles songs on our peg. Four by fours got stuck on hillsides, bonfires burned in icy glades and four generations of family all warmed their hands together.

And once the guns were locked away again, we stayed quite late.

~

The next day, there was another party.

Over the Rainbow

I was starting to flag a little at this point – we all were. But everyone dug deep. Claire had tweaked a muscle doing the splits to Reef's 'Place Your Hands' and Geronimo had mangled his toenail trying to put on the wrong welly when we left the shoot lunch at 2.00 a.m. I wasn't carrying any physical injuries, but the cumulative effects of chronic merriment were starting to tell.

Fortunately, there were footmen assembled at the gatehouses with Bloody Marys, which were passed through the car windows and had perked us all up by the time we'd got to the other end of the drive about ten minutes later.

There were carol singers in the rotunda as we entered, and my family scattered in all directions in a flurry of high fives and hugs.

I was, briefly, alone.

'Isn't this wonderful,' I said to the guy I found myself standing next to. He looked at me quizzically. 'This house. I do really love this house,' I said, adding a bit of detail.

'Wonderful,' he said. 'Wonderful. And you know what the really *clever* thing about this house is?'

'What's that?'

'It's not *ack-tew-wer-lee* a very big house.'

He was stone-cold sober and completely serious. To be clear, it's a Grade 1 listed stately home. It's even got one of those rooms full of heads on plinths and a quince lawn, laid out by Repton.

'Have you had a poke around?' I said. 'Seen the heads?'

'Yes, of course. That's the clever thing, because you see it's really not *all that* big. Not *all that*. It's really clever. Deceptive. Really very manageable.'

Geronimo, nineteen, and his seventeen-year-old twin

Last Christmas

brothers had found the walk-in humidor by the time I caught up with them and were smoking large cigars with the host. They were full of gusto, particularly excited about the DJ they'd clocked back in the rotunda, who was performing later in the nightclub downstairs.

It was a delicious, glittering blur of a day. There was hot roast partridge with bread sauce, and coronation chicken, and Armagnac of several vintages inside, and it was bright on the South Terrace where cocktails were served, beef roasting on a rotisserie over an open fire.

The DJ was not only brilliant, but also kind enough to play 'Rasputin' when asked.

~

By the time I came up the next day, I was really struggling.

But it was Francie's party, and I really like Francie.

We stayed late. Stayed very late, up with her talking about kippers, I think.

~

And then, at last, it was Christmas Eve. We were due to head to a carol service at Dom's house in Adlestrop.

Adlestrop is one of the most beautiful villages in the world. There's a poem about a train stopping there 'unwontedly' in the height of summer, which perfectly evokes the magic of the English countryside.

It's sacrosanct, actually, Adlestrop. It hasn't really changed since that poem was written a hundred years ago.

The railway station has gone, the branch line – which ran from what was Chipping Norton Junction, cross-country to Cheltenham – has been torn up, but the 'Adlestrop' station sign is still there where the platform was, which only adds to the feeling of it being a place of absolute rest.

These places are hard to find in the English countryside. Especially somewhere so close to London. Even our farm has a railway station, the old branch-line terminus at one end and a busy road at the other. The richest people live in the quietest corners, like big friendly spiders with Armagnac webs and butler tentacles.

I really like Dom. He's one of the cleverest people I've ever met. He has a job that he never, ever discusses, employing thousands of people, which he seems to perform effortlessly. I've never seen him so much as look at a phone. His dad, a vicar, started a whole new branch of the Christian Church, and although I've never seen Dom in a church apart from his own, at Christmas he has lots of spiritual gravity and is very much at ease with himself.

I like him for many reasons. And he likes me, too.

'You know, what I like about you, Alex?' he said once.

'I've got absolutely no idea,' I said. 'But you never really do know why people like you, do you?'

I mean, I know why I get invited to all these parties – it's because I used to be in this massive band. But it's true that, as I said to Dom, you never really know why people actually like you.

'Cos you're happy,' he said. 'I can't be dealing with sad people.'

I'm not sure if *happy* would have been the way I'd have

put it when I woke up that morning, but I was giving it a really good go.

The week of cider brandies, cigars and smorgasbord was really starting to take its toll, and my legs were now also sore from the last round of 'Rasputin' with the kids at the party before last, the one with the good DJ.

Nobody wanted to go anywhere, especially me, and yet again, to my continuing surprise, I appeared to be the most sensible person in the room. Possibly, the only one.

'We can't not turn up,' I said, though I needed to curl up with *Chitty Chitty Bang Bang* and a cup of tea and the grannies as much as everyone else. 'Look, we've said we're going so we *have* to make an appearance. Have to. Non-negotiable FACT.'

Geronimo, trooper, agreed to come.

My residual cider brandy levels were off the scale, so I had a couple more after a long, late family lunch, just to get the festive spirit back.

We had trouble finding a driver and arrived a little late. Filling the pews were many faces I didn't recognize and there was a palpable sombre note in the air. I could feel it, and so could Geronimo. As the singing began, I didn't quite glean the sense of communal *joie de vivre* that had decorated the week's gatherings. We even backed off the descant harmonies, shaking our heads at each other as the last verse kicked in.

It wasn't until the guests were filing down the monks' walk from the chapel back to the *schloss* that we learned that Dom's father had passed away and his memorial service had been held in the very same church the very day before.

Nonetheless, in the house there were all kinds of food flying out of the kitchens and an exceptional bar, even by the standards of the week. And when I was introduced to a new drink involving warm cognac, I fully reconnected with the spirit of Christmas. I found the Steinway in the music room and couldn't help thinking that this was the time to give Wham's 'Last Christmas' a go. I was just getting into the second verse when I heard Dom saying, 'Do you need a lift home? I'll take you home, now, Alex.'

I could tell it wasn't because I couldn't remember all the words. It wasn't even midnight and there were still people there. I reassured him that I was literally leaving now, and found Geronimo and told him that we needed to go, and go right now, but I didn't quite know why.

We hadn't planned our exit as carefully as we might have done, so we started walking home, old sore legs and young sore toes.

Fortunately, it was only about three miles home, and fortunatelier still, twenty minutes or so in, when we were about halfway down the drive, someone offered us a lift.

It was a Porsche – a really fast one, but a two-seater. I'm about six three and Geronimo is taller. It took a while, but we worked out that if we took the roof down and he lay across the bit where the golf clubs go with his feet sticking in the air, it would all work just fine, provided that I curled up in a ball in the front seat.

Back at the farm, we popped into the office for a drink and a debrief. It was all a bit fuzzy.

'Did I go man overboard? I mean, I know I do occasionally, but not since "Uptown Funk" at Rita's, and everyone said that was brilliant. And that was ages ago.'

Last Christmas

But I'd definitely been bounced out. The fresh air from the zooming Porsche had done a great deal to reinvigorate me, and the mists began to clear after a couple of espressos.

I had been smoking in the room with the art and the *objets*, as usual. There was quite a lot of religious iconography in that stateroom, relics of the host's dear departed father's calling.

Someone I didn't know had said, 'I'm not sure if you're allowed to smoke in here *ack-tew-wer-lee*,' although I had before, on occasions beyond count.

Nonetheless, to keep the peace, it being Christmas Eve and everything, I'd dutifully picked up a little ashtray on the sideboard where I was leaning and extinguished my cigarette.

Sadly, unlike some staterooms, not quite everything in this one was actually an ashtray. I'd happened upon an antique figurine of the naked baby Jesus and stubbed my fag-end out by rubbing it around in his crotch, quite obliviously.

When I was alerted to what I'd done by the person who'd asked me to stop smoking, I tried to put a positive spin on it.

'Yikes. That is pretty bad. What can I say? I'm a promiscuous iconoclast! I just can't help it, clearly. That's probably why I was so good at being in a band. First rule is break rules. Good job it's solid Carrara marble. Exact same stuff Michelangelo used by the looks of things. The same quarry is still open as a matter of fact.'

I picked baby Jesus back up to demonstrate his robustness by tugging one of his little arms. The arm came off along with plume of impossibly fine, glittering, powdery dust.

I managed to wedge it back on, but I was still rubbing it clean where the cigarette ash was – which was the last place anyone would want it – when the maid spotted me. And she must have snitched.

Geronimo agreed that it would have been that much funnier if Dom's dad hadn't been buried the day before and if he hadn't been a vicar and if it wasn't Jesus' birthday that we were there to remember.

2.

Britpop Trousers

January

By the time I was back in the office, the days had been getting longer for two whole weeks and snowdrops, the first whispers of summer's looming detonation, were sprouting everywhere.

All in all, I was quite looking forward to getting some work done. And yikes, was there work to do. Five kids, a bunch of cats and two hundred acres of exploding farmland is quite a handful.

And this year, there was more.

I've been in a band, exactly the same band with exactly the same people, since I was nineteen. On and off. For eight years it had been off, but completely unexpectedly, a band meeting had been called in London just prior to Christmas. The first one in years. The previous meeting was a car crash that had haunted me daily ever since, and there had been very little communication between us or with management. Until the live rep had called:

'Wembley,' he'd said.

'Yes,' I said. 'Yes.'

I meant, 'Yes, I'll do it', not, 'Yes, I'm listening', but the conversation didn't seem to be over.

'That's the *national* stadium . . .' he went on, slowly, patiently, clearly, like he was talking to a child. Or an idiot. '. . . has made a successful application to Brent Council to stage additional live shows in July.'

He took a breath. For ages.

So I said, 'Yes,' again.

'The FA, that's Wembley's owners, have just signed it all off.'

'Yes?'

'As *such*,' he said, and he paused again, 'there is now an extra slot available. At Wembley. The stadium.

'I think you should all try to talk about it?'

'Hmmmm.'

I wasn't entirely sure what there was to discuss. A discussion sounded quite tricky, in fact. It was a yes or no question, and it looked like a pretty simple 'yes' to me.

~

I'd found my erstwhile copilot and bridge partner drummer Dave 'David' Rowntree, outside on his phone when I got to the hastily convened summit. When he saw me, he moved it away from his ear, shook his head and waved his hand across his throat.

'Don't think it's gonna happen,' he whispered, and looked very sad. I don't know what he'd got wind of but he was pale and despondent, and it was cold, and London was horrible and grey and ugly and smelly.

I stood there fidgeting and smoking, shivering and nauseous, while he wrapped up his call. Then we stood there silently facing each other for a moment, minds whirring. Was it all back on or not?

He shrugged. Then he pushed on the buzzer.

And then there we all were, all four of us. Damon, the singer and the boss, said, 'So what are you doing now?' like it was a job interview. Graham, guitar, said, well, yeah, he was doing a film score for Jez Butterworth, and Dave said, oh, yeah, he was scoring a new show for Disney. And they all had records coming out.

And I said I was, well, basically trying to make a really, really big Frazzle. Really big: a sort of crispy, bacon-flavoured edible plate type of thing, but it was proving much more difficult than I had expected.

Formalities dispensed with, we got down to business. We bashed through a few songs to see how it felt, and it felt good, and less than half an hour later, to my great surprise, that was that.

I was in a band again, and we were going to play at Wembley – that's the *national* stadium.

And we were going to make a new album as well.

Which, once the giddy euphoria had worn off, presented numerous challenges.

That one meeting aside, it had been eight winters since the band had last played a note together. We hadn't spoken to, or even shouted at, each other for two years. And there was no guarantee that anyone else was still interested. That's something you never really know until you actually start trying to sell tickets – and we had a lot of tickets to sell. This would be our biggest ever show. Would anyone really be persuaded to go to downtown Brent on a Saturday night in summertime?

Tickets went on sale just before Christmas. Booking lines opened at 7 p.m. At about five past the phone rang: the whole thing had sold out and the next night was available and we should grab that as well, the agent said.

So that was all much easier than expected – and it was on that note I'd parked everything for Christmas and let my hair down. But now it was January.

~

I live on a farm that has nineteen kitchens, where I make cheese and cider and giant crisp prototypes. I am surrounded by chefs because we spend all year devising a world-beating food festival. So, as explained, I've got my hands full, and often my mouth, as well.

I had become enormously fat. Hadn't, in fact, seen my own scrotum since halfway through lockdown. It was pretty bad: Jim Morrison just before his last bath, only much older.

I had been making up for it by being funnier and kinder in my daily life, but now I had to get in decent shape. No ifs or buts. You can't point a Super Trouper at a lard-arse. Doesn't work.

So, on 2 January, I gave up smoking and stopped drinking and drove straight to Oxford to get some gym shoes at Decathlon. When I finally found the one person who seemed to work in that enormous warehouse full of enormous people – a child – I said:

'I need some running shoes, type-thing.'

'Who for?' he said.

'For me, actually.'

'Oh! Sorry,' he said.

What was perhaps most unsettling, I reflected on the drive home, was the fact that it's usually a successful career in popular music that is the trigger for excess and Fat-Elvis muppet-ward mayhem. My band, never the earliest to bed, was actually going to be the balancing factor here: I would be deploying rock and roll as a form of rehab, which could be tricky.

~

Contemplating going to the gym is quite similar to having a hangover. There's nausea. There's dread. There's a strong inclination to go to bed and watch *Chitty Chitty Bang Bang*.

But the good thing about going to the gym is that, unlike giant Frazzles or mid- to late-career albums, there is no hit and miss about it. It's absolutely guaranteed to work every time. I was going to have to hit the gym hard. I needed professional help.

Fortunately, Claire likes going to the gym and it was agreed that we'd both see a trainer, twice a day, every day, until I fitted into the trousers I wore the last time Blur played together.

Cashflow was tight. Getting a trainer wasn't going to be cheap, but this was a big fat emergency. I hung the Britpop Trousers above the big table in the office as a reminder.

It took a while to track the trousers down. Eventually, I located them in one of the farm's remoter sheds, the Dutch Barn. The Dutch Barn was built between the wars for storing hay. It's vast, and we still use it for storage – it's completely full. I'm not entirely sure what of. Everyone dumps stuff in there. The trousers were between my dad's old inflatable sailing dinghy and two huge crates of roller skates.

A farm is just a collection of sheds, really, and we've grown into them over the years. A lot of them are kitchens now, but there is a gym shed as well.

That is, a freezing-cold shed full of spiders and a grim library of instruments of muscle-tone torture accumulated over twenty years of snacking and comebacking. I went to have a peek inside.

No one had been in there for ages, clearly. The whitewash was all flaking on the mats. The heating was broken, as usual, and there was an ice-crusted puddle on the floor by the radiator, along with a doily of dead flies and half a bottle of champagne on the windowsill.

It was going to be painful.

~

Trousers located and gym inspected, I sat down at my desk and kicked off the business year by calling the trainer, Angela. She's older than Claire and me and a total hardass. I really like her.

A likeable personal trainer is a godsend. You need them to be nice if you're going to be spending hours together, daily – hours of torture. I got fed up with the last one tasked with getting me back in shape for the previous tour when he told me that, if I was hungry, I should drink a nice glass of water – possibly the stupidest suggestion I've ever heard.

I'd taken matters into my own hands at that point and

gone running instead. All over the farm, all over the forest, up and down the valley; through rain, through snow; at dawn to begin with, and then, eventually, in the dead of night. That was like floating in space: bouncing along to the rhythm of my breathing in perfect solitude, a head torch illuminating a bright cone of silent landscape in the pitch-blackness.

I really loved night running, but by the time I was ready for band photography, I'd worn out two pairs of trainers and one pair of spikes and both my big toenails had fallen off. This time, I was too fat to run. I was facing a 2XL next-level fatastrophe, and I couldn't face it on my own.

Fortunately, I had Angela and I had Claire to help. Sort of. Claire is fearsome in the gym. A stalwart of countless epic Help for Heroes bike rides, Everest Base Camps and sprint triathlons, she, like Angela, is a total hard-ass.

~

The long journey back to my old trousers began with a gentle warm-up on the running machine.

'You're too fat to go on there, Alex. You'll break it,' Claire protested.

Angela guffawed, but Claire was getting cross. And she was wearing boxing gloves.

Angela soon got used to us arguing in there. She banned us from talking about work because that was when it tended to get most heated.

But that didn't stop me shouting. And swearing. Often both at the same time. I lose all sense of self-consciousness when I'm in the fat-burning zone. I swear so much and make so much noise I wouldn't be welcome in a public fitness centre. Even running around London parks people used to give me funny looks because of all the huffing and puffing.

Christ. On we soldiered. On and on. Kettlebells, dumbbells, barbells. Pull-ups and push-ups. Weights, cardio and core. I learned to fear any exercise with Russian or Bulgarian in its name, or spider or gorilla or donkey.

Angela would often turn up with a cheap new gym gizmo she'd found on the internet, or a huge tractor tyre, or length of fat rope that she'd found in one of her sheds. We worked every single muscle until it sang, and for the entire first fortnight it was like I'd been squished by a juggernaut.

I did, in fact, break the running machine, much to Claire's annoyance, and then I broke the massage chair, and I felt broken, too, but the programme was working.

~

There are endless diets and endless different exercises, but the weight-loss formula is a very simple one indeed: exercise more and eat less.

Eating less is the worst part.

We'd been to a so-called spa retreat in Thailand many years ago and done a week-long fast, living off nothing but broth and coconut water (with a slice of pineapple on the Wednesday), but going to Thailand wasn't a viable option.

The plan was to broth it, nevertheless. We'd fattened a couple of pigs up for Christmas and the extended household had smashed through numerous roast shoulders, baked hams and a truly extraordinary amount of bacon over the holidays. There was a shed full of ripening salami that would be ready in spring and, more importantly, a chest freezer full of all the best bits: the bones and the trotters.

I was a vegetarian for twenty years. I never thought I'd change, but the sheep that came with the farm were delicious. Sheep are still farmed in the way they have been for hundreds of years: they just nod about, eating the grass – pretty much wild animals, but with healthcare packages and predator protection bolt-ons.

Commercial pig farming has become a lot more industrialized since the farm was built. There's a wood beyond the garden wall where we always keep a couple, though. The old-fashioned way, They truffle around, beyond grateful for all our leftovers and stuff that would otherwise go to waste.

Our biggest success had been with a pair of Gloucester Old Spots that we'd got as piglets. A brewery left behind a whole pallet of mulled cider at Feastival, and there was a pallet of rice left in another field. We soaked the rice in the cider and fed it to the pigs. The meat had a remarkable flavour that went particularly well with Cheddar cheese.

Our latest pair had grown over the past two years into half-tonne whoppers. The meat was like unlike anything I have ever eaten: uniquely satiating. The bones were just brimming with potential. So I was pretty excited by the idea of living on that broth.

I already had the perfect place for making it in mind. There's a lean-to outside the back door of the house where, a few years back, we installed a rotisserie barbecue that my mum had got me for my birthday. It has a big turning spit for joints and tiny twirling skewers for kebabs and it makes everything delicious. From the moment it arrived the kids wanted everything cooked on it, even sprouts, and even when it was raining. Once the barbecue was installed there, the old lean-to quickly filled up with more batterie de cuisine: deep fryers, griddle plate, wok burner, pizza oven: an additional kitchen basically, just outside the kitchen.

Humble stocks are worth dwelling on for a minute or two. They are the gasoline that all Michelin-starred kitchens run on. You can make a more than adequate chicken stock by throwing all your chicken leftovers, even KFC, into your biggest saucepan filling it up with water and leaving it so that it sits just, *just* below boiling overnight. Strain it off in the morning and you're in business.

Having good stock to hand is better than money in the bank. It enhances pretty much everything that comes out of the kitchen and improves day-to-day living immeasurably. We'd rotisseried the whole back leg of a cow on Geronimo's eighteenth the previous year and the bones that eventually emerged from that took five days and the world's biggest cauldron to stockify. The liquor it gave was the heartiest, most nourishing and restorative elixir I'd ever tasted – insanely good with vodka and tomato juice.

Still, even that gear had nothing on the Britpop-trouser stuff I was stewing up now. When you're as hungry as I was, absolutely anything is delicious, but this stuff had its own fan club before it was even ready.

You need think carefully about where to cook it, though, because if you spend five days boiling something indoors, absolutely everything in the house will start to smell like it. It's been more than five years since the Romanian farmhand and his wife left, but the Dairy Cottage still smells of boiled cabbage.

I wired up a griddle plate under the lean-to just outside the back door. In the freezing cold of January, the smell of the developing stock was an arresting, tantalizing treat for the nostrils. It was a great pleasure just to spend time with that smell. Delivery men lingered and asked what it was. Dogs and cats called temporary truces and lay together in the loving arms of the aroma. You could see faces light up as they caught a fragrant whiff. Even the impossible teenage kids and their friends hung around, taking it in.

And when the stock was finally ready, it had such depth, and it had such might, and yet it was still light. Savoury, flavoursome, wholesome, home-grown heaven.

I lived off that, the occasional spinach leaf, and the last of summer's apples and pears.

And I felt good.

~

Before I met Claire, I travelled so much that Dave, the other half of Blur's rhythm section, and I realized the most practical and sensible approach was to complete professional pilot training and buy an aeroplane. He still flies, dive-bombs the farm whenever he's passing, but, apart from Antarctica – occasionally – I never want to go anywhere any more.

With every passing year it becomes harder to tear myself away from my own machinations here. Not because I can't; more because, when the time comes, my heart sinks a little bit as I realize there's something I'd much rather be doing at home.

So, when the day came, I did not, for example, feel all that keen to beetle back to Bournemouth to cut the ribbon on the new 'canteen block' at my old school. But I'd found it hard to say no when they got in contact to ask.

Still, I reflected as I got up at 4 a.m., at least it meant that I didn't have to go to the gym that day, and fortunately, even for someone who was basically living off gruel and coffee, the words 'canteen block' seemed to have an almost magical appetite-suppressing effect.

It was filthy and bleak in the dark. Dawn was obscured by relentless rain and relentless spray all the way due south to the coast.

The headmaster's office was damp and freezing. As I took a seat, all I could think of was the last time I'd been summoned there, forty years ago, for an almighty roasting for drinking wine on a school trip. The furniture was all exactly the same as it had been then, and the walls hadn't been redecorated.

The whole day it didn't stop raining. The entire brigade in the canteen block were delighted to have me. They'd cooked an enormous feast, and my old year group had rallied round and invited some of my old classmates back as a surprise.

It was a surprise party, basically, for me.

I drank a lot of coffee and answered a lot of questions.

The headmaster offered me a glass of wine. I had another coffee. Then, as the bell rang, absolutely everyone was going to the pub, and it was still raining, and I said I'd really love to join them all, but I had to go to my mum's immediately.

I ran away across Bournemouth to discover she had very kindly spent most of the day cooking tea for us, but I explained I had to go home and cook dinner for the kids. I persuaded her to come with me, back to stock and serenity.

There was so much rain and darkness to plough through that it took forever to get home, but a fire was blazing and there was laughter in the kitchen where the kids had taken matters into their own hands. Geronimo was making pizzas and Beatrix had made about a gallon of garlic butter. Sable, her older sister, doesn't like pizza. She was doing some dough balls.

'Dad?' she said. 'My mocks finished today. Can I have a party?'

'Absolutely no fucking way. Let both your grannies be my witness,' I said, looking to them both for support.

'What did he say?' asked my mum. Her hearing aid had run out of battery in the car.

'Sable wants to have a party,' said Nana, Claire's mum, and my mum's face lit up. 'Oh, won't that be lovely.'

Claire and I exchanged glances.

Fifth-form parties are the craziest parties I've ever witnessed. Nothing that ever happened with the band comes close to that level of mayhem. And as for liabilities – it's a total nightmare. At sixteen, you can drink, legally, if supervised in a home setting. And you can have sex, legally, at sixteen.

Trouble is, the last thing sixteen-year-olds want at a party is adult supervision, and the first things they want are, apparently, to try to puke themselves to death – and, failing that, to get to know each other better – and even if it's legal they still need to be protected from their own sheer gusto and exuberance, and it's a nightmare before you even start to consider that not quite everyone in the fifth form is actually sixteen, anyway.

It was an absolute, hard no.

Could she then, Sable said, have just six people for dinner in the thresher barn, no drinking, and we said yes, of course that was fine. And then there were twelve coming, because she'd said they all needed plus ones, and by the evening itself there were twenty-five and Claire had gone into full producer mode and compiled a database of emergency contact numbers and allergies, plus we'd had to pull in the entire Feastival support network: St John's Ambulance, roving security to scoop up anyone who passed out in the garden, the whole nine yards.

The girls arrived first, and I said a cheery 'Hello' to one of them. She actually looked over her shoulder to see who I was talking to, then shrugged and looked at the rest of them as if to say, 'Who does this guy think he is?'

I put a chef's jacket on and remained behind the lines cooking and invisible in plain sight until they'd eaten, and then left her twin brothers Arte and Gali running the bar, while Geronimo DJed – all of them were armed with two-way radios, and instructed to spot and report any pukers to Command HQ in the kitchen, where Claire, Nana and I paced around and Beatrix kept a keen lookout through the window until midnight.

It was a blessed relief to get back in the gym on Sunday morning.

3.

In the Studio

February

February came, with all its snowflakes and all its promises. All white, and then all grey, and then all shades of green, all in the same day.

I just sat there, rapt, gazing at it all whenever I could. I had so many things to do that not doing any of them, just for a minute, was bliss.

Farms don't have brakes fitted, and the pace picks up in February, as trees have to be planted and seeds all sown by spring – which was, on certain mornings, already bear-hugging the whole of England's great green machine back to life from its slumberings.

Momentum was gathering on every front. I'd found a new groove and I was just about, almost, equal to it all. Double daily gym workouts – dread torture just a few weeks previously – had become a regular part of the grind. Muscles fizzing and bones aching, my whole body was sore, but it was noticeably changing shape and I could put my socks and shoes on without making any noises.

But it wasn't all smooth sailing: Monday's farm planning meetings are vital. Having to miss one to go to London was always a kicker, but especially in February, when most of the planning takes place. This February, I was going to miss every single meeting: the entire diary had been cleared for the whole month for recording in London. Which was daunting just by itself. It's hard to think of a band who don't all hate each other thirty-five years in, let alone one that has made a good record at that point.

I'd taken refuge from the prospect of it all in the physical and the metaphysical: staring out of the window, head in the clouds, when I wasn't in the gym.

~

The first Monday of February, the morning alarm went off like a starter's pistol. I went to the gym with Claire, and then we ran through everything in the bath before the day kaleidoscoped into countless spinning plates and wobbling vases.

I woke the kids who, all apart from one, swore at me and went straight back to sleep. I selected two crimson apples and two fat, juicy-looking pears and, with brief satisfaction, slotted them in my bag with my phone, laptop and headphones. As the car rolled up the drive, I felt paradoxically still, like I'd jumped out of an aeroplane and was drifting down to London serene under a parachute.

Of course, it would have been better if it wasn't the pilot who was peacefully parachuting elsewhere, but the whole point of holy matrimony is that there's two of you. Claire was wearing zebra-print salopettes and wellies and she was on the phone, smoking, as she waved goodbye.

Gaz, the driver, said, 'I put a water bottle for ashtray.'

'You'll be pleased to hear I've given up smoking,' I said.

In the Studio

'Wasn't as difficult as I expected.' And I got to work. There were already some emails to deal with and a couple of calls to make, but by the time we'd joined the M40 the far side of Oxford, I'd cleared the decks, put my headphones on and opened a folder entitled B.O.D.

It contained around fifteen mp3s from Damon, the bones of a new record.

When a bunch of musicians go into a studio, they've got to start somewhere, and that's what a demo is: a starting point. What the finished thing might end up like is anyone's guess. All I could really do beforehand was familiarize myself with them a bit and get ready to go with the flow.

So I listened to them all again now, and made some notes as my parachute swish-swashed gently through the Chilterns.

I was absolutely cacking it.

You can consciously prepare making music to some extent, but the important thing is to listen and to react in the moment. It's like falling in love, maybe. The chemistry is either there, buried in the subconscious, or it isn't, and there was nothing I could do about that on the M40, so after a while I moved on to the other music I had to listen to: two new demos from Ruby and Violet, filmed on Violet's phone.

~

I'd known Ruby and Violet Eliot since the day they were born, because their dad, who I met in the filthiest boozer in Soho, was my friend when being the bass player in Blur was my whole life. He died, suddenly, just after they were born, but Claire, who hadn't really known him, really hit it off with their mum. In fact, she and Bianca had become closest friends. There was always a slight whiff of witchcraft about the whole situation, but it really pleased me. Claire would often take all the kids down to visit them in Cornwall during the holidays or meet Bianca in London.

I only really saw Bianca at Feastival. She came every year as Claire's wingman, and it was on the Sunday night after Feastival the previous year, when everyone was kicking back and relaxing, that she asked me if I'd ever heard the girls sing. The office was full of steamrollered chefs and smashed artists dancing to 'Rasputin' at that point, so we went over to the piano in the house.

As a songwriter and producer, I've listened to a lot of music by new and would-be artists over the years, and I reckon I've said about the same thing every single time: 'It's great. Yes, really great. It really is. Well done. Never stop doing it,' because these are people's precious dreams in your hands, and you can't crush them. But the truth is always this: the intro is too long; the verse needs to kick in quicker. What's the tune? Where's the melody? It needs a middle eight. And, what's it called? I mean, I've listened to it, but I still don't know what the song is called. And most important of all: Why? Why? On every level. Why are you doing this? Why does it have to exist? Why should I listen to it?

So that's what I was expecting would happen. I'd be sitting there thinking, why am I listening to this and missing out on all the 'Rasputin' action.

I stared into space as Ruby and Violet composed

themselves at the piano. Then I looked over at them. They're identical twins. Tomboyish. Still at school. They didn't want to be there either. Much less than I did, I suddenly realized.

'Go on,' said Bianca. 'Just one song.'

Ruby glowered at her mum as she struck the opening chord and drew a breath. She began to sing, and her sister sang harmony.

I suppose if you spend enough time in the stream, you're bound to come across the odd golden nugget, and I've encountered a number of brilliant young singers over the years. Lily Allen sang backing vocals on the records I made with her dad while she was still at junior school. Sophie Ellis-Bextor was, I think, still at secondary school the first time I met her, and Florence Welch a student at Camberwell College of Art when she came to the farm and cartwheeled around the garden in hot pants in between vocal takes. But I'd never seen or heard anything like this. Not ever.

There was no doubt about it in my mind: It was a hit song. A heartstopper. 'Superficial Love' it was called.

Of course, no one ever really knows if a song is going to be a hit until it's recorded and released, and there's a million things that can go wrong, but I sat there listening, and the question 'Why – what's the point?' didn't come into it. Their voices had an almost supernatural resonance. There was pain and yearning and longing. They were born to do it.

'They've been singing just like that since they were, ooooh, toddlers,' said Bianca. 'What's it called? Harmony?'

They'd never been in a studio at that point. Since that

evening, they'd been coming to the farm, and I'd given them as much time as I could, trying to help them find their feet in the studio. They could write songs without even thinking about it, but that's just the first half of the battle. Bringing songs to life in the studio is the next step, and it takes a while to learn how studios work.

Initially, they'd written all their songs on a piano, but these new demos were guitar-led. One was called 'Mirrors', and one was called 'Echoes', and they were both strong, and it was those songs that filled my head as I arrived at Thirteen, Damon's studio, and made myself a cup of tea.

~

Often, being the bass player is like being a spectator in a 'who can be latest?' competition between the singer and the guitarist. Dave was already there, fiddling with his drums in the live room. Drummers always have stuff they can fiddle with, but I found a bass I liked in 1995 and I've never had to fiddle with it since, so I went to sit in the control room.

We've always recorded backing tracks with Dave in one room and the rest of us in the other, partly because drummers are even more annoying than usual in studio situations and partly because the only way to get the beat recorded clean, without what everyone else is doing bleeding into the drum mics, is to isolate them.

The control room at Studio Thirteen has one more window than most: a skylight looking up into a grey abyss of West London fumes. The other, more traditional one

In the Studio

looks directly into the live room through soundproofed glazing. It's big, as control rooms go, but it's busy: home to an enormous vintage AMS Neve mixing desk, and many pairs of monitors, which is what loudspeakers are always called in control rooms. There are racks and racks and racks of outboard sound-processing equipment, and every other spare inch of space is taken up with obscure and highly collectible keyboards and drum machines. It was crammed even before Graham's tech had set up his pedal board and half a dozen of his favourite guitars just inside the door.

The tech was waiting just outside.

'All right, Steve?'

We hugged.

No sign of the singer or the guitar player, but James Ford, the producer, was there. I'd never met him before. He was fiddling with a monster 1970s Russian synthesizer that was humming like it was about to explode or catch fire at the very least.

Sam, the engineer, and Giacomo, his assistant, both of whom I'd met briefly before Christmas, were weaving to and fro with jack leads, patch leads and a soldering iron.

I tiptoed through a minefield of bespoke foot pedals around teetering towers of priceless audio gizmos and, clutching my bag, my bass and my brew tight to my body, collapsed delicately into a two-seater sofa on the far side of the room at the back.

'Stormzy's seat,' said Sam.

'Mine now,' I said, claiming the spot.

It was half past ten.

Over the Rainbow

The guitarist and the singer had both arrived by half past eleven.

We started playing by midday.

By the time we broke for lunch we had a master take in the bag. The first demo had transformed from a promising doodle into a fleshed-out, almost finished thing that just needed a few overdubs and backing vocals.

I don't know how it works, but it does. As soon as the four of us start playing together it's good. And it always has been. The world falls away and it's just us in the music and in the moment, the same as it was the very first time we played together, when we wrote a song that we still play now.

It's effortless, it weightless, it's joyous, like riding a bicycle downhill, like flooring it in second gear, or catching a wave or a breeze. Like an engine, a fountain, a fire.

My heart was pounding.

~

'That was easy. How was it easy?' I murmured to Dave, as I savoured my second pear. I was trying to keep the apples for the ride home.

'We've been doing it for thirty-five years,' he said. 'Anything's easy if you do it for thirty-five years.'

~

I don't think I've ever eaten an apple with as much satisfaction as the one I'd saved for the way home: a Spartan, I noticed: a venerable local variety from the Victorian orchard in the walled garden. They must have been eating them on the farm for centuries, just exactly the same as this one, I pondered, and probably during many a February when there wasn't much else to eat.

A brand-new branch of Greggs had just opened on the Peartree Roundabout and I wondered if the sausage, bean and cheese pasties would be hot as I messaged Angela the trainer and Claire to say I'd be back in twenty minutes and would meet them both in the sweary sweat box.

That night, Claire and I collapsed into bed, too sore to even hold hands until the alarm went off again at 6.30 a.m.

~

And that's pretty much how it ran for the next few weeks. I snapped right into the routine: waking up just before the alarm went off at 6.30, groaning in the gym for an hour, followed by the momentary bliss of a hot bath, then into the office for an hour, before taking a running jump out of the flying farm and pulling the London ripcord.

It was a unique blend of complete exhilaration and total exhaustion. The studio felt less like a pressure cooker than usual, more like a place of refuge and stillness. Maybe for all of us.

A lot of the bass was recorded lying down with my feet dangling over the end of that two-seater that Stormzy had spent the previous six months farting on.

It wasn't always like this. When we made our first album, *Leisure*, in 1990–91, it took months and months. Everything was recorded onto massive spools of fantastically expensive tape. You were limited to twenty-four, or, by linking two tape machines together, forty-eight separately recorded tracks – and, ideally, live drums need a dozen or more. There were no computers at all in studios back then. A fully loaded laptop and an affordable rack of gear the size of a suitcase offers more sound processing and editing flexibility than any studio in the world did when we recorded 'There's No Other Way'. The editing on that was done with a razor blade.

The way most records are made has changed: samples, loops, plug-ins, autotune. In fact, you can emulate so much there's barely a need for session musicians in the modern studio, let alone a band. Bands are unwieldy and hard to manage because, by and large, drummers are mad, guitar players are moody, singers are mental and bass players are just an anachronistic luxury. So music tends to be compiled rather than captured in the 21st century.

Bands are difficult and expensive to run. Like ageing stately homes. And the record business ran out of money a decade ago. The industry just can't afford to make them any more.

But making music with a group of people that have been doing it together for years is a very precious thing. Music is, always, a collaborative process. No one can do it on their own. Not even Prince.

There is nothing that can touch the sound made by a close-knit group of people who have been playing together for years and years and years, playing as though their lives depended on it. For many years, all our lives did, and actually, I'd suddenly realized before lunch on that first day, they did still.

In the Studio

~

The recording process was a case of capturing something that happened in that room, with everyone playing live; capturing something special that after eight years of soul-sucking hurt and frustration was suddenly giving me strength.

As was the precious pork life-support potion which, by the end of February, we were eking out with dashi – that is, stock made from seaweed. As we all know, if you're hungry enough pretty much anything is delicious, but dashi genuinely is. It's the very rails that Japan runs on, and I love Japanese cuisine above all others.

Apart from in Sumo wrestling for entertainment purposes, Japan doesn't tolerate fatties at all. Apparently, if you grow overweight in Japan, you have to go and see your boss to discuss an action plan. If you remain too chunky then you have to go to see his boss. If that doesn't work, you would then technically have to see his boss but that has, as far as anyone knows, never happened.

Making dashi is one of those things, like baking bread or frying an onion, that is guaranteed to draw people to the kitchen saying, 'What's that?'

Kombu, which is sometimes the sole ingredient, is nothing more nor less than dried kelp, but it imparts a flavour as bottomless as the ocean it grows in, and it's absolutely bursting with complex salts that give the savoury hit the Japanese call umami and that underpins their whole cuisine.

The day the rhythm tracks were completed I got home and slid an A4 sheet of the stuff into the stockpot like an oversized teabag and suddenly, magically, as it rose to a purring simmer, the old lean-to smelled like a three-star Michelin restaurant.

I'm not saying I didn't harbour wistful, forlorn dreams of smashing a KFC, but the winter brassicas and alliums cooked in, and eaten with, a blend of the pork stock and ocean soup, followed by the very last of the summer's fruit roasted by the fire in the kitchen, windows rattling with all the worst that February could throw at them, was fine dining.

The album was taking shape, and I'd lost two stone.

I felt, not for the first time, that good fortune was on my side.

4.
Half-Term

February

The cat situation was fully out of control. There were about eight of them and I had no idea how or why we had so many. And it was getting busier and busier in the office. There were quite a few kids around, too, as they were all on half-term, but most of them sleep all day and rarely want to speak to me anyway.

The diary was filling up daily with more and more Blur stuff, but now the backing tracks on the new album were – pretty much – all in the can, there was, mercifully, a brief pause. This came partly because this is always a good point to step back and take a minute, and partly because everyone in the band had other commitments.

Mine was Feastival. It needed my full attention.

Claire and I got to work, making the sixty-yard commute from the farmhouse to the office in the thresher barn, trying not to step on any of the cats.

The farmhouse itself is a crooked Escher maze of tiny rooms built for tiny people who must have been very, very chilly in the winter. We've had to install heating and insulation, knock out walls, raise the lintels on all the doors and build an extension on the kitchen so everyone can sit down to eat at the same time. There is, however, a church-like scale to the thresher barn, and something of the monastery about it. It's calming.

In the early days, the 'office' was located in the only warm spot in the house, right next to the fire. As our first farming endeavour – cheesemaking – and the number of children grew, we had to move the office out of the house and into a converted stable. That office doubled as a control room for a very basic recording studio. Then, as Feastival grew, we needed more and more space and more help. It was a proud day, around ten years ago, when we moved everything into the thresher barn. The thresher is now part office, part development kitchen, part studio and part bar.

People often say, 'Music *and* cheese! That's unusual!'

'Ah,' I reply. 'Monks have been doing it for yonks.'

In fact, pretty much everything we do here, monks and nuns have been doing for millennia. Even throughout the Dark Ages, while it was all mayhem and carnage out there, in the perfect isolation of their walled sanctuaries the brothers and the sisters peacefully devoted themselves to growing food, brewing, singing and spreading the Good News – which is pretty much what Claire and I do now.

Feastival, which takes place at summer's zenith, the August bank holiday weekend, is where all those elements come together all at once, right here on the farm. It takes all year to plan and promote. And it's a lot: 25,000 people, 10,000 tents, 30 stages. Countless bars and food stalls. Ballerinas and boffins, tightropes and trapezes. There's always work to be done, decisions to be made, problems to be solved, fantasy Frazzles to fry. Or maybe bake. I was still working on that.

Half-Term

It brings my family together. Even when they were toddlers, the kids absolutely loved it. Their excitement as slime factories and their favourite stars were booked, their unbridled glee throughout the build and the weekend itself, were a big part of my own motivation.

This summer, all five of them would be teenagers – and teenagers are hard to connect with. Impossible, sometimes. I'd found the best way to get the kids to talk to me was to drive them and cook for them and give them all summer jobs working at, and helping to organize, Feastival. But this summer, I was going to be away from the farm and the family every single weekend.

Kids need a cook and a driver and someone to get them in and out of bed. They don't really need an absent musician, playing in a band that didn't mean anything to them.

When Blur last played Hyde Park, Geronimo was still in short trousers. It was a perfect summer evening and during the last song, 'The Universal', the entire crowd was as one: arms aloft, ecstatic, singing. Many crying, all exhilarated: an entire Royal Park jammed to the nines with a united crowd sharing a moment they'd maybe never forget.

In the two-beat pause before the final instrumental chorus, a silent, spinning, split-second of sonic climax, I'd stolen a glance over to the side of the stage where I knew the kids were watching, to see what they were making of it all. They were all under the monitor console asleep and oblivious. A cute puddle of bare arms and legs and brightly coloured ear defenders.

The big moment for them all that summer had been when the Lego arrived. Lego were a sponsor of Feastival that year. From their bedroom window, the twins spotted,

Over the Rainbow

dangling from a crane in the Front Field, a full-sized shipping container painted Lego yellow with Lego branding on the side and made to look exactly like a huge Lego brick. There was a heart-stopping commotion that drew the entire family running to their room above the kitchen.

The twins were jumping up and down on their beds pointing and shouting, 'LOOK! LOOK! LOOK!' But it was obvious because even at a distance the docking Lego-brick mother ship was about all that was visible. The other three instantly went hysterical and that is about the happiest I can ever remember our children.

When you're organizing a festival or any kind of party, that's kind of where the bar is. You're aiming to delight. Even if it's just a neighbour coming over for a beer on a sunny afternoon, you want to get the glasses in the freezer beforehand so when you pour the beer it's close to zero and you get a mist on the glass. It's not brain science, it's just attention to detail: that's what makes the difference. As the party gets bigger, there's a lot more scope to ramp up the excitement, but you can't lose sight of the detail. That's what takes time and effort.

As I scanned the big calendar to get a bearing on the coming weeks, I noticed I was drumming my fingers in mid-air, not even on the table, and that Claire was looking hard at me because I was jabbering away to myself.

~

I'm constantly telling the kids: 'Just try to do ONE THING and try to do it really well. Trust me. The rewards are endless, and you'll never come unstuck.' Before they were born, I had pretty much devoted myself to playing the bass: four strings, one note at a time. That was it.

I was going to struggle to follow my own advice this year. The diary read like a laundry list of global hedonistic extravaganzas of the early 21st century: a modern-day grand tour of massive knees-ups.

Following the Wembley announcement, more offers had come in, and Blur had been booked to play about a dozen of the world's biggest and best festivals. We'd be playing to – literally – millions of people, and then it was all back to mine for a big old party.

I couldn't work out whether I was the luckiest man alive or whether I had bitten off more than I could chew.

On paper, Feastival was actually shaping up pretty nicely. The headliners were booked. That's always the first big headache. 'Early bird' ticket sales were strong, too, even before we'd made any line-up announcements. But there are a thousand festivals up and down Britain every single summer, and the only ones that survive and thrive are the ones that offer something unique.

In many ways, the stupider you can be when you're trying to plan an event of this size, the better. The kids were really good at coming up with crazy ideas when they were young. The mud-pie kitchen was a stroke of genius that sat very well alongside all the Michelin geekery. I'd been trying to get health and safety to sign off a zipwire from the giant redwood tree behind the cooking stage directly to the backstage VIP area. I was brimming with ideas, but having ideas is easy. Making them happen is harder.

And it was all grey, and all wet, and it was hard to even imagine summer. Plus, there were more immediate pressing concerns. With a sigh, I turned my attention to the top of the day's pile: the ongoing drainage clusterfuck in Railway Field.

~

This farm was, right up until 1932, a mere speck within a vast fiefdom: the Sarsden Estate.

There is immense wealth in the Cotswolds today. A lot of poverty, too – but that isn't always as obvious as all the helicopters. There's new money, there's dynastic wealth, there's Norman blood money. There's cash riches, asset riches, the whole damn embarrassment – but I chanced upon a map of the Sarsden Estate at the local heritage centre and the imperial scale of it beggars belief.

Thousands upon thousands of acres; many, many entire villages. Maybe a hundred square miles in all. The architectural masterpiece of the squire's stately home nestling, invisible, in Repton parkland, was serviced by its own Brunel railway terminus *and* branch line just down at the bottom of the valley.

The old railway hotel and the cattle auctions next door were all sold off when the estate was broken up in the early thirties, at the same time as our farm and dozens of others, along with the stately home, all the villages, quarries, Iron Age forts, tumuli and everything else.

And that's when this whole drain problem started.

All the drainage from the old railway hotel (now an old folks' home) and the old cattle auctions (now a mix of

Half-Term

businesses and homes) runs underneath Railway Field. This worked fine historically, because the drains were built to last, and everything for a hundred square miles was owned by one person, so if anything did ever go wrong, it was that person's problem.

Now, dozens of different owners were participating in 'historic drainage rights' and although those drains were indeed built to last, they were built in the 1800s and nothing lasts forever.

We're not talking storm drains or field drains here. We're talking drain drains. Trust me: when they stop working, you can immediately tell without looking what kind of drain you're dealing with.

I'm pretty sure that none of the small businesses, auction house residents or old folk had the merest inkling of where their drains even went or what their legal position vis-à-vis historic drainage rights might be until the old clay Victorian ones started to fail.

First thing I knew about it was when I was walking down to the station and was confronted with a burbling sea of human slurry.

It was a total shocker. Railway is one of my favourite fields. An endless carpet of buttercups, bordered by woodland and ancient hedgerows, it is perfectly punctuated dead centre with a natural pond surrounded by Domesday oaks and brimming with wild watercress: a tableau of seclusion.

It took a while to work out quite what had happened and where it was all coming from.

We cleared the field up, but the stuff just kept on coming, and after a couple of days I couldn't see for shit or solicitors.

I hired the go-to drainage guy, the leading leak litigator in the land. He had confidence. He had experience. He quoted case studies, court battles. The sheer scale of his knowledge – his very particular expertise in this, his chosen specialist subject – shone like a star. True expertise, in anything – anything at all – is the most precious of all treasures. So at least I had that to marvel at.

Trouble was, the old folks' home, the businesses, the residents, all had their expert litigators too. It was an almighty intellectual scrum.

The case folder was a foot high at this stage. Even giving me the top line took half an hour on Zoom. It was still a gigantic mess and it all needed sorting before August, because you can't throw a party on an open sewer.

We got off the phone and I went outside to shout and swear at the cherry trees.

~

It's hard enough running an independent festival as it is, because a huge proportion of the UK's major festivals are run by one group. It's a bit like the way lots of the Indian restaurants in this country operate. There are loads of curry houses, and they all look independent, and in some respects they are, but they practically all buy all their ingredients from one vast invisible supplier, lurking in the background. Same with most festivals. They're called different things, but it's all the same content. You'll see the same side shows at a festival in Suffolk in June as you will in Cornwall in August – it just moves around.

So it's vital to do new stuff. Stuff no one else has

thought of. That's the fun bit, and the hard bit, and the bit that brings true satisfaction. But if there's one thing I've learned, it's that you just never know what's going to work until you try it.

Sometimes there are lucky accidents. The very first year we had a sponsor drop out at the very last minute, right in the middle of the build, in fact, so where there was supposed to be an 'immersive brand experience' there was absolutely nothing.

We'd just taken in all the hay so we made an enormous haystack and put it in the empty space. The headliners, KT Tunstall and The Kooks, went down well, but, my word, it was that haystack that people really, really loved. It was absolutely mobbed, all weekend long: children and adults alike. By Sunday night it was just a huge pile of hay, which on Monday morning turned out to contain about six hundred quid in loose change and half a dozen iPhones.

It was a serendipitous reminder that the simple pleasures are so often actually the best, and also tend to have the most universal appeal.

Universal appeal – along with good drainage – is absolutely key to making a successful family event, but it's pretty hard to make a whole family happy all at the same time and almost impossible with teenagers. There are only two situations that are one hundred per cent guaranteed to do the trick with my lot.

One is immersion in water: bubble baths, hot tubs, ice baths, steam rooms, pools, the sea and so on. Never fails.

The other is Scott's restaurant in Mayfair.

I'd already booked Mark Hix for Feastival that year (in fact, I'd promised him the twins' bedroom for the weekend).

He pretty much wrote the menu at Scott's. We also had a good line in to an outfit that were going to set up a load of wood-fired hot tubs in the ancient woodland (the farm has about eight acres of it).

But when I told the kids, they weren't impressed. I still had some work to do, clearly.

~

I tried to get Railway Field out of my head, and focus instead on the question at hand: how to get Lego-level delight at Feastival that year. It's the simple, elemental things that people connect with in the countryside, I mused. Sunshine being the big one. Sunshine is, in fact, the ultimate big one. I mean, food only grows at all because we live on a planet that goes around a star, which, by the time you've parked your car at Tesco and found a blessed pound coin and a trolley, can be very easy to forget.

Whatever perspective you're coming at it from, weather makes or breaks the entire weekend. When it comes to an outdoor event, the weather forecast becomes a terrifying thing.

If the sun's shining, everyone is happy with a pile of hay. If it's raining and especially if it's windy *and* raining, ABBA could be playing a surprise comeback show and everyone would still be miserable. That first year I did wonder if I was actually mad spending twelve months a year on a business that lives or dies by the weather until I realized that's just what all farmers have done forever.

I couldn't count on sunshine, but I could count on my fail-safe, back-up, get-out-of-jail card: the Cheese Hub.

Half-Term

~

The Cheese Hub is a double-decker sheep shed I built with the spare cash last time Blur got back together. Each August, it transforms into an indoor disco with multiple bars and restaurants. Even in a hurricane we can ensure the party doesn't stop.

The Cheese Hub evolved organically because, as I've mentioned, you just don't know what's going to work until you try it. It started life as a tent in the corner of Front Field, home to the main arena. It was conceived as a place where wine experts would bring the world's finest wines and pair them with the world's finest cheeses and it seemed like things couldn't get much better. It wasn't until year three or four that Claire suggested we play music in the Cheese Hub as well. I took her advice, more to keep her quiet than anything, rigged my phone up to the big monitors from the studio and set up shop in the corner of the marquee.

Nothing could have prepared me for what came next. As soon as I got up there, a crowd formed that didn't stop singing and dancing. I DJed for two solid days and nights. When I was there playing records, the place was completely rammed; the minute I stopped, the whole place emptied.

It literally couldn't fail. It was totally bombproof. I played the nastiest songs I could think of, crimes against taste from the far reaches of YouTube, on low bandwidth. They sang every word.

By Sunday afternoon I couldn't speak. I was utterly spent. But no one else was done singing and dancing: not by a long shot. We radioed Jo Whiley, who was hosting the main stage, to come and lend a hand. The tent had just hit fever pitch when she had to go back to introduce Mark Ronson's set. I could barely stand and it looked like the whole thing was going to nosedive back into cheese and wine when Geronimo said, 'Dad. I've got this.'

I remember croaking, 'Are you sure? It's not actually as easy as it looks, son,' as I handed him the mic and my phone. I was kippered: voiceless, delirious, done.

Five minutes later I was crying tears of joy, along with everyone else in there. Geronimo James, thirteen years old, was rocking the Cheese Hub like it had never, ever rocked before.

That was the moment when the Cheese Hub proper was born.

He just knew exactly what to do. 'Are you ready?' he said, offering the mic in the crowd's direction.

'Yeah,' they replied.

'I said, ARE! YOU! READYYYYYY?????' he hollered, and as they all screamed assent, he hit them with Avicii and it was all over. For the next two hours, he worked them into a frenzy. It was magical. They were all going at full 'Rasputin' when I joined him up there for a father and son 'Double Decker', taking it in turns to play songs, to finish the weekend.

We could hear them shouting for more all the way back to the house. And then it was wonderfully peaceful and perfectly still in the kitchen, just the two of us. I basked in the sweet satisfaction of a year's work paying off, brimming over with fatherly pride.

Geronimo was very quiet and pale, though. He wouldn't

tell me what the matter was, said he needed to speak to his mum, who fortunately arrived at that exact point with his brothers and Mark Hix.

Claire took Geronimo up to his bedroom. They were gone for a while.

'We need to call a medic,' Claire said, over the walkie talkie.

'What the hell is it?'

'It's his testicles. He's in agony.'

I was terrified and radioed it in immediately on the emergency channel.

His eleven-year-old brothers caught wind of what was happening, and the fact they found it the most amusing thing they'd ever heard was no solace.

'GERONIMO'S GOT TO SHOW THE DOCTOR HIS BALLS,' they yelped, and ran up to their bedroom to tell Hixy and anyone else they could find.

An onsite medic arrived, and I accompanied him to Geronimo's bedroom. Poor child had been the king of the world twenty minutes earlier.

The medic had a wild beard and a restful manner. 'So, tell me what *you've* been doing,' he said, sitting on Geronimo's bed, hands together as if in prayer.

'DJing,' said Geronimo, a faint smile breaking through the pain on his face.

'How long for?'

'Three or four hours.'

'Were you jumping up and down?'

He nodded.

'He was, actually. We both were,' I added.

'I see,' said the medic. 'And did you drink anything?'

Over the Rainbow

'He's thirteen,' I said.

'Anything at all?' said the medic.

'Just the Red Bull,' said the boy.

'How much Red Bull, about?'

'Only about one. Just that one, er . . . case.'

The medic frowned and then cracked a smile.

'Well, if you drink a case of Red Bull and jump up and down for four hours, this is pretty much what you might expect. You'll be fine. Try not to overdo it in future.'

Very good advice.

But it's been boom and bust in the Cheese Hub from day one.

Disaster lurks wherever triumph reigns.

~

I was alone in the office, and nobody bothered me for five glorious minutes, which I spent staring out of the window at the apricot blossoms.

Then I called my favourite YouTuber, the best-selling author and mathematical genius Matt Parker.

'Matt,' I said. 'Can we make a time machine in the Cheese Hub this year?'

5.

Easter, Full Beluga

March, April

From that stationary start in January I'd hit warp factor nine, gone supersonic, full battle stations before the daffodils were over.

The grooves on the album – working title: *The Ballad of Darren* – were all in the bag, so my work was done and pretty much dusted, but I still had to turn up to the studio, just in case anything needed reworking or completely starting again, which can always happen.

Daily, even before I'd arrived in West London, I was up to my neck in WhatsApp groups and six-way Zoom gangbangs with Feastival's teams: PR and marketing departments, ops crews, teachers, accountants and stars of YouTube. Emails, texts and missed calls snowballed as the business day rolled on. But it's beyond bad manners to even look at a phone in a control room, let alone, heaven forbid, to pull out a laptop; and an outright insult, an act of treason, not to turn up at all.

The only thing that would have been worse than not turning up at that stage would have been for me to say anything or try to contribute helpful suggestions to the proceedings whatsoever.

All I am required to do in Blur is play the bass, and it was an exquisite pleasure to sit there quietly, even on Stormzy's slightly uncomfortable sofa, and witness the band's unexpected rebirth unfold before my ears day by day, but I was exquisitely hamstrung.

It would have been *contre l'esprit* not to eat my apple with everyone else while they tucked into their takeaways at lunch, but it wasn't like I was having lunch. I was specifically not having lunch, not till I could get my Britpop bags on. Still, I had to turn up to the sessions and to lunch. Succeeding really is ninety per cent turning up or at least turning up reasonably well prepared, but there was more and more stuff to prepare for and turn up to as each day went by.

I could barely remember what cheese tasted like, yet I spoke to cheesemakers in the car, talked to chefs about menus and cooking demos, and accountants talked cashflow and drain lawyers talked overflow.

And this was the crucial time for plotting and planning with everyone on the farm, because Easter was fast approaching, and Easter is the hard deadline for planting, felling and fiddling about in the woods.

~

Whatever happens with Blur, with Feastival, or with me, even, the farm is the family fortress. Stewardship of the farm was something I couldn't let slip. But, like in any situation, if you're not putting time and effort in, it all goes wrong – fast. The farm has a trillion tiny moving parts that need a steady hand to shape them.

Easter, Full Beluga

Claire and I had spent a lot of time talking, by which I mean arguing briefly in between sets of burpees, despite Angela's protestations, about trees.

The best day to plant a tree is, as any farmer will tell you, yesterday, and failing that, today.

The farm is home to a good number of towering oaks, and there's a ten-acre sliver of what was Wychwood Forest: the ancient woodland that once cloaked and carpeted the entire environs. The owner-before-last had felled and logged about sixty acres of it, some of which we were doing our best to re-establish, alongside taking good care of what remained.

The helpful farmer who said you should have planted a tree yesterday would also tell you that you plant trees not for yourself but for your children, but I'm not so sure. I've found the business uniquely gratifying.

The first time Blur got back together, in 2009, we splurged all the spare cash we could muster at planting more woodland which had by now blossomed thirteen times and exponentialized into a new world all of its own: a mix of broadleaf varieties that was home to songbirds, owls, voles, moles, toadstools, weasels and wildflowers.

Much as I loved the woodland, it was the fruit trees that made me happiest, and we had planted as many of those as we could. We'd banged an orchard in pretty much every corner of the farm: apples, pears, papples; plums, gages, damsons; cherries, mulberries, quince and medlar were all thriving. Peaches, nectarines, apricots, figs, and even olives, were working to a certain extent, too, and there was one ancient vine in the walled garden, but the grapes never fully ripened. Grapes were, in fact about the only thing that didn't seem to work on the farm.

I'd often wondered if it was entirely sensible planting all those fruit trees, or if it was, at best, an indulgence. But I figured some people had boats, some people had nice cars or season tickets for The Arsenal. I had orchards.

Actually, Claire had orchard fever much worse than I did. She seemed to spend March constantly on the phone to the Brogdale Collections, the national fruit tree archive, tracking down strange fruit and obscure varieties.

I'd been trying to rein her in a bit, in fact, but as we bickered through our twice daily April gym sessions with Angela mediating, it began to look like she might be bang on the money.

That fruit was currency.

All the classic British tipples – the monastery favourites like mead, perry and cider brandy that had been overlooked and undervalued for decades – were, very suddenly, in high demand.

~

The more I travel, the more it becomes apparent that, wherever I am in the world, someone has worked out how to convert whatever grows there most prolifically into highly concentrated alcohol, and market it all around the globe.

So, you get vodka from potatoes in Russia, rum from sugar cane in the Caribbean, whisky from Scottish grain. Armagnac and Cognac from French grapes, Calvados from French apples. The Romanian couple who had bequeathed a lasting scent of cabbage to the Dairy Cottage had a line in to a simply sensational no-label plum hooch from back East. Even the grannies got stuck right into that gear and would ask if there was any left every Christmas. We'd had to take matters into our own hands since they'd left – hence the large number of plum trees.

Plums had always been my absolute favourite, anyway. There is nothing to touch plucking a soft warm plum from the branch and eating it right there, fresh, in the glow of high summer.

Along with their bright country cousins, the damsons and the gages, plums grow here like wildfire. Massed cascading droplets, bough-bending frozen fountains of crimson, green and gold: summer's echo of fully bedecked and be-baubled Christmas trees.

However, plum hooch was something of a growth market. It was English wine, and particularly English sparkling wine, that was having a moment. And while the plums were going like the clappers here, grapes just didn't quite fly.

I found this enormously frustrating for various reasons. Partly because, despite being a poster boy for British cheese, if I had to pick a specialist subject on *Mastermind*, it probably wouldn't be cheese, it would more likely be sparkling wine. Blur's original incarnation in the 90s had been one long bubbly bonanza. I'd done my champagne research.

When I was a kid, the nation was still wrestling with the hard fact that food ran out during the Second World War. Cooking skills weren't passed on to my parents' generation because there wasn't anything to cook.

Bacon was still rationed when my dad was well into his twenties. As a result, recipes were lost, eating habits changed, and a proud and noble food culture vanished. A tiny island whose sheep and cattle breeds had shaped the eating habits of the entire world was viewed from Europe as the home of bad weather and bad food. But that was all changing.

Even in the 1950s, when food was still rationed or in short supply, Ian Fleming conjured yearning for fine English cuisine as fantasy in the Bond novels. As the Great British Food Machine lurched and lumbered back to life, Elizabeth David's and Jane Grigson's books brought back forgotten recipes and eating habits to a ravenous people. Fanny Craddock and Delia Smith held the nation's rapt attention on their television shows, but the game really picked up when Marco Pierre White became the first British chef – and the youngest chef ever – to win three Michelin stars in the mid 1990s. It picked up further still when he gave them back to Michelin saying something along the lines of, 'actually, you're not fit to judge me'.

That was a turning point, and the tide was now in full flow, all across the board – the cheeseboard. Not least, Britain had gone from producing a single variety of cheese in the war years to just under a thousand different types at the last count.

British wine was having an even more exceptional rise to eminence. There was no English wine commercially available at all prior to 1990. Winemaking was the preserve of hobbyists and amateurs. As I write, the best English sparkling wines rival the finest in the world.

Easter, Full Beluga

~

Eating and drinking well is a kind of superglue that brings families, friends, colleagues – the whole of civilization – together.

We've come a long way in a short time, but as a nation the French are in a completely different league when it comes to enjoying the pleasures of the table. This is an entire people who are willing to buy fresh bread every morning *and* every evening *and* take a three-hour break in the middle of the day to buy cheese and go home to cook lunch.

Moving to the farm had, in some sense, been a twenty-year lunch break, so I can relate to that, but, still, the level of culinary competence and devotion to good eating of the average French person is staggering. I went to France as a teenager and never wanted to come back. I studied French at university because I fell in love with a culture that revolves around eating and drinking well. Everyone in France eats from the same menu. They are a nation of gourmands. And they are the undisputed grand masters of high-end food marketing. Champagne, in particular, is a case study in marketing genius.

For hundreds of years the *grandes maisons de champagne* have built their Cristals, Krugs and Dom Pérignons into some of the most coveted brands on the planet. There's not a single situation that wouldn't be improved by the stuff if you believe the marketing: from breakfast to business deals. Even the cheapest champagne evokes a sense of opulence, luxury and celebration. In the 1700s, champagne producers were gifting the stuff to royalty so the dear impressionable bourgeoisie would get FOMO and want to buy some.

Over many generations, over centuries, the champagne houses invested vast fortunes associating champagne with the biggest celebrations and the best times. On victory podiums it's sprayed at television cameras. How did they make that happen? They sold us the very air in the bubbles. The corks kept popping and the world couldn't get enough.

Thing is, sparkling wine was most likely invented by English monks. Although the French monk Dom Pérignon tends to get the credit, it's now thought that the art of secondary fermentation in the bottle – the science behind the fizz – was originally perfected in England.

If you leave unpasteurized apple juice in a wooden barrel over the winter, by spring you'll have a barrel of cider. The apples come with everything you need. The natural ambient yeasts on the skins ferment the natural sugars in the juice and, hey presto, abracadabra, cider, cheers. That's the most basic recipe, of course, and the key advance in cider making came alongside developments in English glassmaking. It was found that adding a dose of sugar to the cider as it was bottled made it fizzier and stronger. Forget sliced bread, bottled cider was a true paradigm shift: a double dose of fermentation made it twice as good. It was so good that cider was literally currency back then. Farm workers were paid in the stuff, so whoever had the best cider had a pretty keen advantage.

The coal-fired furnaces adopted in English glass manufacturing meant that only England had glass that was robust enough to withstand the high pressure generated by all those miraculous bubbles.

Easter, Full Beluga

Grapes turn into wine in the same way apples turn into cider, and the monks in the northern reaches of France's grape country soon cottoned on to the merits of stout glass bottles. Their slightly lacklustre still wines were transformed by a spoonful of sugar.

Indeed, they made a virtue of the fact their wine needed sugar adding to it.

And there's no doubt, it's delicious stuff. Light and refreshing and particularly good with food, especially anything cloying, like cheese or chocolate. Bubbles act as a palate cleanser so that the last mouthful always tastes as good as the first.

Years back, I'd trademarked the name Britpop because I thought it sounded much more like something I'd like to drink than to listen to. After years of playing around with cider recipes, we'd launched Britpop Cider the previous summer, and it had outsold everything else at Feastival put together. Now, ahead of what we hoped would be a vintage summer of Britpop music, I really wanted a sparkling wine I could call my own.

Only problem was, I didn't have any grapes. They were literally the only fruit I didn't have lots and lots of.

But I kept turning up to work.

And eventually it became clear that a lack of grapes wasn't necessarily a problem.

~

We'd been in touch with Furleigh Estate, a sparkling wine maker in Dorset, where I grew up. They said they might be able to help and sent a case of samples to the farm.

Over the Rainbow

I hadn't been drinking at all. There just hadn't been time. But when the crate of Furleigh's finest arrived one sunny morning a couple of weeks before Easter, it was clear these were extenuating circumstances.

We called in a dozen more English sparkling wines for comparison, plus a few ringers from France, and set up a blind tasting in the office the day the kids broke up for the Easter holidays.

I've done a lot of taste tests over the years, initially in cheese competitions, and later on as a food writer in just about everything, from mince pies to mead. Going in blind is vital. It's strange, but as soon as you know which one is the expensive one, or the hip one, or the one you thought you'd like the most, or want to like most, it all gets completely skewed. You have to judge everything that's put in front of you with no preconceptions.

With cheese, there are three qualities to consider: appearance, texture and taste. With smoked salmon it's actually just two: taste and texture. That's because the colour of smoked salmon subconsciously affects the way we perceive its taste. Exhaustive studies have shown that adding orange food colouring somehow makes us think it tastes better; even experts can't help preferring the orangest salmon. Professional salmon tasting is all done in darkened rooms under ultraviolet light that makes it all look the same colour. A very serious business.

As was this.

I'd wanted a sparkling wine with my name on it more than I can ever remember wanting a Brit Award with my name on it, and now, after two decades grunting with fruti-culture, it looked like one might be about to land in my lap.

Easter, Full Beluga

Word had got out about the taste test, and there were more people than usual in the office when I got there, still panting from the gym.

'Flava' Dave had laid out every piece of glassware he could find on the big table and an excited, chattering babble of everyone from bookkeeper to bee man was offering to assist.

People always seem to want to help Flava Dave. He draws a crowd. He used to be the chef at Gifford's Circus, a Cotswold institution that his wife ran, but when she left, he did too. He came to work here, in the market garden. He initially came for a part-time job, as he wanted to get more of a grip on home-grown ingredients, but it was instantly obvious to us both that we needed each other. Within next to no time, he was growing all the food for and overseeing the restaurants in the Cheese Hub over Feastival weekend. He's a very good man to have around – knows his alliums. He acted as sommelier for the mystery taste tour.

Blind tasting champagne was trickier than other wine tastings, because I wanted to get a good look at the mousse, that is, the pleasing carpet of foam that forms on top of the liquid as it's poured into the glass. Hiding the bottle in a carrier bag with just the spout sticking out, Dave decanted a dozen mouthfuls into Great Aunt Sylvia's old crystal glasses. That seemed to do the trick.

The sun burst through the skylight and the bubbles fizzed in liquid gold. We said 'Cheers!' and the games began.

I could spot the non-vintage Tattinger a mile off, all biscuits and brewer's yeast. There were a couple of clunkers: one that tasted like pear drops and one like perfume, but other than that, they ranged from pretty good to outstanding. And absolutely everyone had something important to say about absolutely everything we tasted.

It's funny, but I've never seen a big bust-up at a food competition. I was a judge on the Costa Book Awards once. Nobody on the panel listened to anybody else. Everyone was convinced they were the cleverest person there and only their opinion mattered and the whole thing was an ugly squabble. It was never like that at the British Cheese Awards, or the Great Taste Awards, or the BBC Food and Farming Awards. I made lifelong friends on all those gigs. There is something about eating and drinking together that creates consonance and harmony.

And by the time we were halfway through, seven or eight bottles in, it was all getting very harmonious indeed. Geronimo was having the best day of his life, and by the time we'd slugged our way through to the finish line he'd found and plugged in a microphone so the winner could be declared to an increasingly boisterous crowd.

And there was a clear favourite, by unanimous consent. It was 'light and bright,' said the bee man. 'Bales of summer hay,' said Claire. My notes said, 'Peaches tongue jacuzzi. YEAH.'

It was the Furleigh gear.

Geronimo cued up 'Rasputin' and the bookkeeper was going at it with the builders and I called Furleigh then and there and said we'd take the whole lot.

Then I realized it was already ten o'clock. I jumped straight in the back of the car because it was a busy day in the studio: our first day back after the break.

Easter, Full Beluga

~

When I arrived at the studio, I was sent directly upstairs to yet another room of vintage keyboards and drum machines that also contained a medical examiner, who was waiting to see me. I hadn't had a medical for years, but it was an insurance requirement for the tour.

The guy asked me how I was feeling. Much as I was tempted to say, 'Frankly, part terrified and part hungover because I had sixteen different types of champagne for breakfast,' I just said, 'Yeah, fine. Pretty good, actually.'

He prodded and he poked and he pricked and he looked and he listened and he sent me back downstairs.

There was a palpable sense of relief in the control room, maybe partly because we'd all been poked and prodded and none of us had been sent straight to hospital, but mainly because James declared that all the backing tracks were, speaking in his opinion as producer, 'fucking cooking'.

You can never tell how good a track is going to be until you've got a finished vocal on it, but even at this stage, it all felt incredibly promising. Damon wanted to really get stuck into the vocals in Devon, on his own, so we thought we'd spend the afternoon listening back to everything to see what we'd done.

It sounded like the best record we'd ever made.

It was nothing short of miraculous. Four people who had barely spoken for almost a decade seamlessly slotting back together.

We decided to listen to it all again, and as evening fell and more corks popped, Geronimo called to say he was in

London. When I told him what I was doing he begged to come along.

I said I'd have to ask, but before the song we were listening to had finished he was knocking on the door of the control room and he was with Ruby and Violet.

The control room of a recording studio is the holy of holies. Even the drummer is not allowed in there most of the time. No record company, no publishers, no WAGs. No one.

Ever since the recording process had started, Geronimo had been asking me endless questions about the microphones we were using, what sort of outboard compression Sam preferred and which version of Pro Tools – questions I struggled to answer, and he could clearly no longer contain himself.

Dave and Graham had left by now and Damon and I were enjoying a post-medical tequila or two, but Damon was, understandably, slightly taken aback by the intrusion.

I introduced them all – Geronimo was in short trousers last time he met Damon – and they begged to listen to just one song.

And they loved it, and so we listened to the entire thing with them, and suddenly the girls were singing, and Damon was looking at me and saying, 'Where the hell did you find these two?'

~

I woke up very groggy back at the farm the next morning, and braced myself for a packed day of Feastival PR. As I blearily posed for photos, I reflected that, although the rest

Easter, Full Beluga

of the day was going to be painful, things were actually shaping up OK – farm, Feastival and band, all, somehow, in pretty good shape.

I hadn't had a day off, or even had lunch, all year, and I was going to be travelling a lot over the next few months. Even if the teenagers didn't want to speak to me in between me driving them to parties – they usually just locked themselves in their bedrooms – it never made things better if I wasn't there. Forget Feastival, the farm and the final Blur hurrah. The twins' birthday was coming up. We needed family time.

But cashflow was tight. Lockdown had hit us really hard. We'd poured our hearts and souls into the farm throughout it, and were reaping the spiritual benefits of our dedication, but not one of the family hustles had worked for a whole year. No gigs, no Feastival, no independent delis or restaurants open to sell cheese in. Things had picked up again, but we'd lost an entire year's income, and three years later we were still digging ourselves out of that hole.

Blur's tour was going to deliver a decent bounce, for sure, but that wasn't why any of us were doing it. The very last thing any of us wanted to do was to milk it. As soon as that starts happening, it's spoiled forever, and it was far too precious to spoil. That was one thing none of us have ever disagreed on. Possibly the only thing. I've been very rich, and I've been very poor, and rich is definitely better – but the first thing you learn when you're rich is just how many things there are that are more important than money. Blur was one of them – and the reunion was actually playing havoc with my cashflow.

Over the Rainbow

I wasn't going to get paid for the Blur album until ninety days after it was delivered, or for the gigs until after we'd done them, and the first one was still a couple of months away. In the meantime, we somehow had to pay for the road crew, production rehearsals, warm-up shows and the production itself, so the band had absolutely no spare cash.

I had enough in the bank to pay the farm and office staff, just, and yes, I'd just blown a couple of grand on rare apple trees, but Claire said we really needed them.

The bank wouldn't help me. I sent them the production budgets. I sent them the bottom line. I sent them the contracts and the fully executed insurance policy with the beefed-up force majeure and inclement weather add-ons that said in black and white, and in plain English, and in Latin legalese, that even if none of it happened at all we would still get paid and they said, no, sorry they couldn't help.

The mortgage on the farm was with the bank as well. I said there must be some headroom there. Look at all these fruit trees we're planting, I said. Look at all the value we're adding. Still it was a hard no from those stuffed shirts. So I went to see Nana to give her a tickle.

Nana, Claire's mum, lives in the shippon, the old dairy, with Julio, her dog. She runs a guinea pig 'stud' in the garden shed, and I love her with all my heart.

'Kath,' I said. 'We're totally skint. Only till summer, but I want to take the kids on holiday. I really need to take the kids on holiday. And probably Claire as well.'

She said, 'You know, Sylvia just left me some money?'

And indeed I did. Nana's Great Aunt Sylvia used to stay with us every Christmas, and had died the year before,

Easter, Full Beluga

having willed her sizeable fortune between every member of Claire's extended family, even some that she'd never met, except Claire. Maybe because she thought we didn't need any money. But when you've got five kids, you always do.

Kath wrote me a cheque there and then so we would have some spending money. I ran back to the house waving it and asked the kids where they wanted to go for lunch to celebrate the twins' birthday.

The response was unanimous.

~

I've known Gordana, the maître d' at Scott's since the 1990s. She ran the Groucho Club when I was doing all my champagne research.

She only really got cross me with me once in all those years: when I tried to ride Roland Rivron's bicycle down the stairs, and fell over the banister onto the piano. But that was all water under the bridge now.

I called her and said, 'Gordy, can you do seven? It's a special occasion. The kids want to go full fruwee.'

A number of my favourite restaurants only take bookings for parties of up to six, but, as I said, she's a dear old mate, and there's seven of us.

'Are you going to behave?' she said.

'I'll be fine,' I said. 'But I am the sensible one in the family.'

We walked to the station, took the train to Paddington and, with my appetite sharpening closer to surgical with every stride, strolled with great purpose through Hyde Park towards Mayfair.

Over the Rainbow

It was springtime and everything was all right.

My heart was beating faster at the thought of going to town on that menu. I knew it off by heart and had been through it in my head on countless occasions since the start of the year. And now, it was time.

I was wondering exactly how they'd be presenting the spring greens as we crossed Berkeley Square and the phone rang. It was Sian, the bookkeeper.

'Yikes. What is it?' My heart was really thumping now. I thought she was going to tell me there was another leak and we'd need to spend the lunch money on lawyers.

'"Vindaloo" has come through again!' she said. 'I've just had the royalties through from Sony. It's earned more than the whole Blur catalogue. Knock yourselves out!'

'Vindaloo' is a football song I wrote twenty-five years ago. Having a football song is half as good as having a Christmas song, because there's only a big football tournament every other year, and it's a bit like having a permanent bet placed on England. The better they do, the more the song gets listened to.

I knew it had done all right because England had a good run in the World Cup the previous year. I'd looked on iTunes right after the round of sixteen and it was number one – at least, a new version of it was. I'd showed the kids, and said, 'Look, Daddy's number one.' One of them pointed at the screen and said, 'Ooh, you're number nine as well.' That was the old version.

I've literally never once heard that song on the radio, and it's not like they're listening to it in Scotland, or anywhere else in the world, but it was clearly still going strong.

Sian told me the number. Royalties are accounted

Easter, Full Beluga

quarterly, and you never know when they're going to land or how much you're going to get, but this was crazy.

I hung up and said, '"Vindaloo" has come through. Fuck it. We're going full beluga.'

We marched up Brook Street, singing 'Vindaloo' all the way to the cigar shop opposite Scott's. The boys decided they wanted cigars, and I thought if ever there was a time for a fat Cuban, it was today.

We sashayed through the line of chauffeurs waiting in their Bentleys and Rollers outside, high-fived Sean, the long-serving doorman, and were shown straight to the very best table, which had been magically extended to seat the whole family.

The 'fruits de mer, to share' at Scott's – or 'the full fruwee' as it is known among the Jameses – is a gastronomic work of art. A true masterpiece. It comes with six different forks, spoons and spanners, and is served on a gleaming silver tower three storeys high, not including the sauces, which all slide in underneath the bottom salver.

As it was their birthday, I'd promised the twins caviar-but-not-the-fucking-beluga. But this had now become a certified full-fruits full-beluga mission, and I was beyond delighted when they decided to get one oscietra and one beluga, and share and compare them. Beatrix asked the waiter if she was allowed to start with a cheeseboard and I almost shed a tear of joy. Claire had the wine list and was deep in conference with the sommelier who was back with a chilled Dom Pérignon rosé while I was still choosing my sides.

What a lunch. I repaired to the terrace for a cider brandy and a cigar with the boys after the mains and told the girls to order all the desserts, plus the petits fours and maybe the chocolates and another cheeseboard, some single espressos and another cider brandy, a large one, and then we all sat down again to discuss where we'd all like to go on holiday with the 'Vindaloo' money. That was when Geronimo's chair broke, but it wasn't his fault and they got him another one, no problem.

It was a perfect moment. Everyone in the family fully satisfied. Lunch had lasted all afternoon. The pre-theatre crowd were already onto their mains. Contentment reigned, no doubt briefly, but nonetheless supremely.

We sat there, dreaming aloud.

Then Beatrix said, 'Can we go to New York?'

And everyone else said, 'Can we? Can we?'

'Pleeeease can we?'

6.

Ten Hours in New York

May

Somehow, we'd managed to keep the new album a secret. Although I hadn't managed to keep it a secret at all. That was absolutely impossible. What was I supposed to tell the kids? They wanted to know where I was going every day, and after Geronimo came to the playback with Ruby and Violet, he was so excited he couldn't help telling his brothers, and then Arte said he hadn't told anyone at all apart from Olly at school, but only because Olly was a massive Blur fan and he was coming to Wembley with his whole family so he couldn't exactly not tell him, and it was getting more and more precarious by the day.

At least the record was in the bag, bar the singing – and that was Damon and Graham's problem. Rehearsals for the live shows started in a week, but that was a whole other business. I'd been sent a list of about a hundred old songs to look at, but I figured there was time for a quick New York raid in the meantime.

~

Over the Rainbow

We stayed on the Lower East Side. It's the last remaining bit of Manhattan Island that hasn't been gentrified: no Soho House. No designer shops. Maybe some art studios, but absolutely no art galleries. Just twenty-four hours of non-stop grime and grunge. It's the complete opposite of the Cotswolds. There's literally nothing that isn't ugly. There's no rest, no silence and no peace, whatsoever. I absolutely love it. Always have.

We'd booked rooms at the Sixty LES, 'The Lezzer', as usual: a blackwashed concrete cigarette carton plonked in the middle of two non-stop traffic jams, right where the motorway meets Manhattan's grid. It's half-hotel and half-nightclub.

The party was just getting started when we checked in at midnight on the Tuesday. It was hard to hear the receptionist over the disco, but it was peaceful way upstairs, and there was a nice note, some nasty cheese and a huge tub of tiny crackers from the general manager in my and the boys' room.

The girls were exhausted and still – just – young enough to be delighted by the minibar, which admittedly amounted to a small convenience store inside their wardrobe. The boys were less easily pleased. They were raring for pizza. Heaven knows, I was, too. New York pizza is right up there with the Scott's fruwee – one of the great culinary marvels of the world.

We all agreed. All we needed was pizza. We'd just grab a couple of pies and then it'd be straight to sleep as tomorrow was a big day.

So, the boys and I struck out into the neon night. There was a dense fug of marijuana fumes in the concrete corridor. It was even fuggier in the lift. The boys were giving each other looks, trying not to catch my eye.

Reception smelled of vomit. We tiptoed round a big puddle of that and made it onto the street where the weedy whiff kicked straight back in.

The pizza place was right on the corner, next to a brand-new non-stop bong emporium.

'What's that smell, Dad?' said Arte, as we joined the queue.

'That is New York pizza,' I said. 'A dollar a slice! A whole pie for ten! How do they do that?'

'No, the other smell, Dad,' said Gali.

'Don't know what you're talking about.'

'Come on, Dad!'

'What's that smell?'

'Smells funny!'

'Yeah, really funny . . . Dad!'

They were all chipping in like that when the guy in front turned to face us. He was huge, absolutely huge, American huge. Hugely overweight and proportionately overheight with huge infinity eyes. He said, 'Man, tha's jus' Noo Yawk Siddy. Y'all wont some bud?'

'No, no we're fine,' I said. 'Thank you. Really.'

He put his hand in the bulging dustbin liner he was carrying and pulled out a handful of sticky green buds.

'C'mon, Daaad! Daddy needs a nugg. Am I right, fellas?'

'Look, I haven't touched that shit for years. Not for a very, very long time. Not since I was about their age, in fact.'

He asked us where we were all from and the boys said England, just arrived, pleased to meet you, delighted. How do you do.

He cracked a grin, pulled up the flap on my breast pocket and poured a fistful of little nuggets right in.

'Welcome t'the siddy, you boys,' he said, tapping my pocket and high-fiving them all in turn.

They literally thought they had actually died and really gone to heaven. We'd been in Manhattan for less than thirty minutes. Pizza was a dollar and drugs were free and, apparently, compulsory.

We ordered pepperoni pizzas, one pie for the boys and one for the girls, as planned, but with that chance encounter, the evening's whole agenda had taken – as it always promises to in Manhattan – a completely new turn.

So, while the boys were taking both pizzas upstairs for the girls, I was making friendly enquiries with the concierge in the Lezzer's freshly disinfected reception.

'Could you please, would you mind, just so that I am absolutely certain where we stand,' I began, 'explain exactly what the score is with smok—'

'I'm sorry, sir. Absolutely not. NO SMOKING ANYWHERE IN THE HOTEL,' he bellowed, looking daggers at me and then all smiles over my shoulder at the next in line: an exceptionally, tall and thin girl with a broad and beautiful face.

'ER! YEAH, yeah, I did hear about that already, but thanks,' I added, wrestling back his attention. 'But what I wanted to ask was: what's the score with smoking weed in the city?'

'Oh, sure!' He brightened up again, unceremoniously dropping the impossibly fragile creature behind me and giving me his full gaze once again. 'That's absolutely fine, sir.'

'But. Like. Where?'

'Oh, anywhere. Sure, that's absolutely fine.'

And then it was the sunflower's turn to ask a stupid question.

The family holiday was off to a flying start.

The towering flower joined me and the boys outside. Yes, sure she had some papers, and she told us where all the cool places were, and we all sat there giggling in the steam and the stench and the sirens and suddenly we had lots of new friends and it was only 2.00 a.m. when the boys remembered how hungry they were and there was a highly recommended Chinese four blocks over and we hit it.

The menu was like a telephone directory and sea slugs seemed to be the house speciality so we got those. We got chicken feet. We got hooves and heads and horns, we went full guts and gizzards, and we all got very frightened when they arrived at the table and then no one could stop laughing when the slugs were presented so we got the bill and ran away back to the hotel to watch *Minions*, the movie.

It was half past four in the morning and Beatrix had woken up, jet-lagged and slightly sad.

She confided after some encouragement that she'd got confused between Miami and New York. She'd said she wanted to go to New York because she wanted to see my friend Robert again, and now she realized it was Miami that she meant. That's where he lives.

She was really sad. She'd come to New York by mistake.

Robert is my best friend. He was best man at our wedding and the kids all love him, too. We speak often, but I hadn't seen him since lockdown.

'Ah,' I said, 'but remember: I met Robert in New York,

and you just never know who you're going to meet here. I'm sure we'll see him soon. Come on! Let's go shopping.'

There was an all-night pharmacy that took up three whole storeys of an entire city block and we stayed there for hours, dawdling then dancing to the music in the aisles, just the two of us – we were the only ones there, a trolley each, stocking up on all the good stuff, just like Christmas, only better: crazy cereals, sacks of crisps, monster marshmallows and all the bonkers candy that Robert brings every time he comes to visit.

We called him down in Miami, woke him up to say we were really missing him, and he said that he'd seen Blur were playing in Denmark and that he might try to come, as his wife, Bea's godmother, is Danish.

'Christ, I haven't even looked that far ahead,' I said. 'Brilliant. When is it?'

'June thirtieth,' he said.

'DAAAAD! That's my birthday! Please can I come, please, please, please?' said Bea.

'I'm going to be working. We'll have to see.'

I wasn't sure how bringing my family to work would go down with the others. Damon had been incredibly gracious with Geronimo – and with Ruby and Violet – in the studio when we'd listened back to the album. I didn't want to push my luck.

But what the hell. It was her birthday.

I said, 'Do you know what? I don't really see why not. Maybe you should all come.'

She told me I was the best dad in the world, and I think she meant it, and 'Ain't Nobody' came on the tannoy, and we sang along, and we danced around, and we twirled our trolleys, and jumped up and down, shopping until breakfast time.

Katz, my favourite deli, right round the corner from the Lezzer, was just opening as we passed. That was the spot to grab breakfast. We thought we'd get it to take away, and eat it back in the hotel.

Katz is a New York institution. They've been knocking out pastrami sandwiches, salt beef stackers, these incredible savoury potato donut things, all sorts, for about a hundred years. I don't think it's had a makeover in all that time, but it's totally authentic, completely genuine. I was explaining to Bea that that was exactly what I loved about the Lower East Side – it was still edgy, right? Hadn't been ponced up and gentrified and made just like everywhere else – when there was a knock on the big glass window.

'Oooh, look! It's James,' I said.

I have five children and six jobs so I don't have many friends. Just Robert, really. And James. James is my other favourite person in the world. He writes films and lives at the top of the hill on the next-door farm. I can literally see his house as I'm writing this and it's the only house that I can see.

'That's so weird,' he laughed. 'I know this is your favourite spot, so every time I walk past, I look in the window. And then, there you were. All right, trouble?' he added to Bea, giving her a squeeze.

He said he was staying right opposite the Soho House that had just opened round the corner, because his new apartment, also just around the corner, was nearly ready, and he had to keep an eye on the builders. Plus he was going to pop in and see Danny, who was opening an art gallery on the next block as well, and they were really old mates, had been ever since Eton.

'I reckon we got here in the nick of time. I'm so glad you said New York, even if you did mean Miami, Bea,' I said, as we wobbled back to the safety of the twentieth floor, fully laden with all the shopping and New York's finest breakfast.

Claire and Sable were sitting up tucking into room-service breakfast in bed. It looked like they may have accidentally over-ordered, but the boys were still awake and ravenous.

And an hour later, by group consensus, we'd made our way to the nearest, nastiest nail bar.

As I pushed all the buttons on the clapped-out reclining massage chair's remote and kind hands tended to my aching feet, I looked to either side of me with sweet satisfaction. We'd taken over the whole parlour, and it was silent as the city raged outside.

Although I'd been busy, I'd been leading quite an insular, even isolated life. I'd been fully consumed every last minute: with work, with the gym, with endless to-do lists. Now, suddenly, I wasn't. I was released. I was in New York and I was with the people I love the most and they were all, for the moment, content. A moment of perfect serenity like I hadn't known all year, and I'd had to come to the busiest city in the world to find it.

7.
Showtime

June

After just four nights in Manhattan – and a good kip on the plane home – the farm looked completely different. I felt strong. The baton had risen on summer's overture and the band was primed to start swinging it again. The first show was just a couple of weeks away. It was time for rehearsals.

Blur have never rehearsed much, partly because rehearsals are horrible and rehearsal studios were always horrible sheds in horrible parts of town, and partly because we've never really needed to. All those years of playing together, plus the fact we all play with other musicians outside of the band, helps. In fact, playing with other musicians is vital. The last band I'd played with was Chic, and there had been no rehearsals for that at all, so I was quite relaxed.

Bottom line is this. Two things matter. Number one: turn up. Number two: play the hits. (As the host of a music festival, it surprised me how often bands failed to adhere to these two basic principles.) Everything else was just nice to have, really. But, of course, we wanted to make these shows as nice as we could.

Over the Rainbow

The Wembley dates would be our biggest ever and, as shows get bigger, there's an inevitable tendency for the band to get bigger, too. Back in the beginning it was just the four of us and our mate 'Nutty' Nev driving the VW Transporter. With the second album, there was a keyboard player as well. Made sense.

By the third album we had a brass section and also an actor, Phil Daniels, doing a cameo. A string quartet was the logical next step for the baroque embellishments of the fourth album, by which time the Mayor of London was also making occasional guest appearances.

Then came backing singers – an entire gospel choir at one point. Percussion, too, and I swear I spotted someone twanging a gourd in one of the quieter moments last time out.

But with age and experience comes the blessed realization that simplicity is art's crowning glory.

We all agreed stripping it right back to where it all started, no smoke, no mirrors, was the best plan. It was right back to basics.

~

There was a lot of ground to cover. I was up at dawn, which lasts for hours in June – up with the birds and the ripening cherries and strawberries to refamiliarize myself with the 'backlist' before going to the gym. I hadn't played any of the songs for eight years – and some of them for over thirty – but, miraculously, muscle memory kicked in.

It was almost a subconscious act. If I tried to work out what the hell my fingers were doing, it was like looking down from a tightrope and I'd fall off the groove immediately, but as soon as I closed my eyes and relaxed it was like floating in the boundless ocean. Bliss. I'd get to the new stuff later.

~

When Blur very first started making records, back in 1990, record sales were what generated the money that made the whole music business merry-go-round go round. It's easy to forget that not so long ago the only way to hear your favourite song was to wait for it to come on the radio or to go to a thing called a record shop and spend four pounds on a single or four times that on an album.

Back then, recording studios were glamorous places with pretty girls on reception. Brasseries. Negronis. It seemed every upmarket *endroit* had to have one and there was always a new one opening: Chelsea, Bloomsbury, Primrose Hill. Literally the only thing – the one single thing – in St John's Wood apart from grand ambassadorial residences was Abbey Road Studios.

On the other hand, there were only a tiny handful of production rehearsal studios, and they were all situated within a five-minute walk of Pentonville Prison, round the back of King's Cross Station.

And then the internet happened, and it all completely flipped. Stepping into a recording studio in 2023 conjured reminiscences of the fast-disappearing Lower East Side: all bare black walls and fugs.

By contrast, the rehearsal studio we were booked into had an entire wall of glass overlooking a pretty canal, but unfortunately it was in Wimbledon. It would have been less bother commuting to Manhattan than to schlep across London in peak traffic, but we all dug in.

The plan was to reveal the fact that we'd made a new record on the day of the first warm-up show, which was taking place in a tiny church in Colchester. Really tiny, but we needed to be ready.

So we all fought our different ways to Wimbledon for the best part of a fortnight. We played the crowd-pleasers. We tried new tempos. We tried new keys. We tried new arrangements, but everything just sounded best the way it always has. It was all ready, already. Still, you can't know that until you've tried everything else.

There was the whole of the new album to tackle as well. But we'd get round to that.

~

Gradually, a giddy sense of travelling began to build, in space and in time: hours in the car daily, followed by hours playing songs that snapped me right back to the moment we wrote them, reconnecting me instantly with the forgotten distant past. There were familiar faces, too. Most of Blur's production team have been with us for years: from the lighting and sound crews to the stage techs. We had spent years and years travelling the world together and living daily side-by-side.

It was hard to believe that all the band and all the gear had ever fitted into Nutty Nev's VW.

Dave has many drums and racks of boxes that flash and all sorts of digital gizmos and triggers. Damon has many keyboards including an actual piano, and Graham has a whole warehouse full of guitars and another one full of effects pedals, most of which he was bringing everywhere with us. They all needed technical support, for sure. But I've got one guitar, and it is completely indestructible. I wasn't sure I needed my own roadie just to put it the right way up, or whatever he was going to do, but it turned out that wasn't up to me, and that was when I was introduced to Richie.

He asked if I wanted a cup of tea, and I said that would be lovely.

As soon as I finished it, he asked me if I wanted another one, and then he found me some cigarettes and an ashtray, and I began to wonder how I'd ever managed without him.

He took my guitar away and when he gave it back twenty minutes later it played differently, sweetly. It looked good. It sounded incredible and he always seemed to know exactly when I wanted a cup of tea.

~

A rehearsal room is a very different environment from a studio control room. There's no intimacy because there's a lot more noise – you literally need a microphone to communicate – and a lot more people. The number of people in the room was, in fact, doubling roughly every forty-eight hours. It was surprising how many people none of us had seen for ages, for years, happened to be 'literally just passing' a canal in the arse end of SW18 and thought they'd pop in to say hello.

And then there were all the people who were actually working with us: at first one, and later two stills photographers, tasked with capturing candid shots for digital and print media, which necessitated two stylists, two groomers and an attendant slew of PR ballers, management execs and their aides. We shot twenty-four magazine covers in ten minutes and then the documentary film crew arrived. Lights, cameras, furry sausage, producers, director, and more execs and underlings.

By the time we started rehearsing songs from the new album, which was called *The Ballad of Darren*, it was so busy in the kitchen that the catering team had to decamp to a Portakabin and Richie had to get us our own private kettle.

~

For those couple of weeks of rehearsals, the gym was horrible, and the travel was relentless, but playing the bass in a rock band, in my band – well, I could literally do that without thinking. I think it's probably the same for the others. Maybe it's one of those things that is either completely impossible or completely effortless. Like floating. I could do it on no sleep standing on my head, shaking my rejuvenated ass and smoking a cigarette while Richie brought me cups of tea all day long.

Which was lucky, because the first gig, straight after rehearsals ended, in the tiny church in League Two Colchester City, turned out to be a global news event. This, despite the fact that the church wasn't quite as big as the rehearsal room, and there were fewer people there than on the last day of rehearsals.

Showtime

It must have been one of the smallest gigs we've ever done. It felt just like going back to the very beginning. There was a strict capacity because of the ancient building's limited number of fire exits, so the place was at full capacity, but still half empty, basically. Damon could address the entire audience without the need for a microphone.

We opened with 'St Charles Square'. The first track on the newly announced album. And that was it. We were back. All back together. No nerves. Pure abandon. Absolute enjoyment.

If I'd been a fan, that would have been the show I'd have wanted to go to most. People that I know, people that I love, people that I admire and respect all came, and they all went batshit crazy.

~

I'm not sure what time I got to bed, but the next day the phone woke me up and it was Dave, my allocated security guy.

He said it was three o'clock and we had to go to Eastbourne now. I asked him what day it was, and he said Saturday. And that was a relief because the show wasn't until Sunday.

But I still had to get there to meet Claire and the kids. All the other Blur kids – Damon and Graham's daughters – were coming to the Eastbourne show, so I'd given my lot the all-clear, too, as long as they stayed out of the way. Claire would bring the younger ones down from home, and Geronimo was coming over from college in Brighton with Ruby and Violet and some of his mates, plus Bianca

Over the Rainbow

and Fenton the ballerina were on their way, as were the Wilson sisters – Ruby and Violet's godmothers – and quite a few of the Feastival team. It looked like it was going to be a good weekend, so I ordered a Bloody Mary and fired up the Bluetooth boombox.

To be honest, I was just a little bit miffed about having a security guy. Not my decision. Smoggy, Blur's head of security, had stipulated it and sprung him on me, unexpected, the previous evening, just before the show.

'Now, what you've got to remember about Alex, Dave,' said Smog, 'is Alex is always, always fine.' Which, I thought was kind, considering how long we've known each other. 'But he is an absolute fucking machine,' he went on. 'You've got to watch him.'

He was talking about me like I wasn't there, and I didn't really want to be. I didn't need this mollycoddling.

'He put fucking Lemmy to bed. Do you remember that, Al?' he said, bringing me back into the conversation. 'In the lift? In Japan? Lemmy? Lemmy pissed himself.'

'Well, it wasn't a competition.'

'It was! That's exactly what it was. And he started it! And you matched him, drink for drink and he pissed himself. In the fucking lift. I was there. Don't you remember? Lemmy? Lemmy from Motörhead?'

'Well, of course I can't remember that, Smog.'

'Pleased to meet you then, Alex,' said Dave, six foot three and eighteen stone of pure tattooed muscle – but he looked slightly apprehensive.

~

Showtime

Leaving Colchester, I realized that this was the first afternoon I'd had off for about a hundred years. It was a beautiful summer's day and we were off to the seaside.

Briefly, I had no responsibilities. I passed my phone to Dave in the front of the car, said, 'Tell 'em I'm busy,' and stretched out in the back with a cigarette, a small bottle of cider brandy and a big smile. Afterglow.

Blur were back! Back! Back!

I plugged my headphones into my laptop and punched 'Matt Parker' into YouTube, and all pretty England sailed past.

By the time we got to Eastbourne. I'd drunk all the brandy and figured out how the time machine would work.

~

The Grand Hotel was one of what were once four 'five-star' hotels on the South Coast, throughout the golden years of the Great British Seaside Holiday. My granddad had run the kitchen at The Royal Bath, the snazzy one in Bournemouth, but since then that superschloss had gone seriously downhill. The Grand, though, had clung on to its grandeur.

Some destination it was. A big old wedding cake basking in the open arms of its bay, all neat lawns and symmetries. Perfectly decorated with topiary, with Bentleys, Rollers and Rangers arranged neatly all along the carriage drive.

And it was lively.

Over the Rainbow

I hopped out of the car and was arrested by the warmth of late afternoon sunshine, the call of the sea. There was some kind of festival pulsing further along the promenade, a band playing up by the pier, and the seafront thronged with happy strollers and dawdlers.

I lit a cigarette, closed my eyes, turned my face up to the sun and was only jolted out of blissful reverie when the band started playing a Blur song, 'Charmless Man', which none of the others wanted to play any more.

'Hey, I said we should do this one, Dave,' I protested, opening my eyes, but he was wafting away top-hatted doormen and disappearing through the revolving doors with my bags.

Then Phil Daniels appeared with his family. We embraced and did just a little dance to 'Charmless Man' in the sunshine.

'All right, Al?' he said. 'Fancy a quick scoop then?'

'Gotta have a sauna and a kip,' I said. 'Family coming.'

'Riiiight. Right.' He chuckled. 'We'll be in the bar anyway.'

And then Dave was back with the room key.

'It's nice,' he said. 'You've got the presidential suite.' He handed me the keys and my phone. 'Do you want this back yet? Claire called, by the way. They're not far off so I'll stay here and greet them and bring them up.

'Oh,' he added, 'and Bournemouth got a result.'

It was a good day.

~

When I got back from the spa, Claire, the twins and Beatrix had arrived, and already made themselves at home beneath the distant corniced ceilings. Startlingly so.

The boys had set up a Bluetooth speaker and were letting it rip with Biggie in their bedroom.

Bea was on the house phone behind a huge desk in some kind of Oval Office scenario, holding a room-service menu, trying to order cheese. Claire emerged from what must have been a bathroom in the fluffy robe and the fluffy slippers.

'Where's Sable?' I said.

'At home, revising,' said Claire. 'Didn't fancy it. Who's that Dave?'

'Yeah, who's Dave?' they all wanted to know.

'Uh,' I said. 'Management clearly think I need a babysitter. Waste of time.'

'What do you mean, babysitter?'

'You know, like a fucking nanny.'

'But he's really nice.'

'Yeah, yeah, whatever. Come on, let's go to the beach.'

'Can Dave come? He's really nice.'

'Oh, Jesus. All right. See where he is.'

I called him. He said he was waiting just outside our door. And then he said word had got out and there were quite a few people outside, and it might be easier if we used the service lift and went out through the goods entrance; and he'd spoken to the manager and got a passkey for the lift, and come this way.

And by the time he arrived at the water's edge, bearing ice creams for all of us as we paddled in the lapping sea, I was starting to think he might have his uses after all and the kids were all completely in love with him.

He taught Bea some self-defence moves which she immediately began putting to good use on her brothers, and none of them could wait for the next 'Code 117', which, Dave explained, was security-speak for an entrance or exit to avoid crowds via a service route.

~

The kids all disappeared somewhere with Dave as soon as they woke up in the morning, leaving Claire and me to enjoy the silent perfection of a sunny Sunday morning.

I headed to the venue for soundcheck, and by the time that was over, it was getting quite lively there, so I snuck back to the hotel for the crowning glory of the perfect weekend – a late afternoon nap.

The party had already started there as well. Geronimo had arrived with Ruby and Violet and a gang of his mates from college. He was DJing off his phone and it was kicking off in the boys' bedroom. Fenton was dancing.

Claire, Bianca and the Wilson sisters, both art professors, were tucking into a case of Britpop in the oval office and spilling out onto the main veranda.

'This is amazing, this fizz – are you having one?'

'Thanks, I won't, actually. Got to work later.'

And off I crept, to the master bedroom where I fell asleep immediately and woke up smiling two hours later, ready for anything.

The presidential suite was packed, and Dave led the procession to the venue on foot in the evening sunshine.

Showtime

Bea and the twins had wanted to do a 117, and everyone was asking what a 117 was, but Dave said there was no need as the fans were all at the venue now.

~

I couldn't even get in the dressing room. The whole backstage was over-abounding with well-wishers. VIPs and stars of stage and screen were spilling out of the production entrance into the loading area outside, where Richie had set up our kettle in a little nest of flight cases. He was waiting with tea and Frazzles and a setlist, just in case I wanted to have a look, and he had to talk quite loud because everywhere from backstage to the bars, front of house to the foyer, was whirring and buzzing with tense excitement.

Dave appeared. 'Smog said you're doing a 118, tonight. Is that right? I'll get a driver to bring your car round.'

A fire exit banged open, and Fenton and the twins tumbled out laughing.

'Fenton got us SHOTS, Dad!' said Arte, and then went into some complicated pat-a-cake routine with her. The big pirouette finish they'd clearly been rehearsing was too much for Arte who collapsed on the floor, giggling.

'Fuckssake, Fenton. They've got school in the morning.'

'Oh yeah. Schoooool!!!!' said Gali, hugging me: holding on like a child and grinning like an idiot.

'Jesus. Where's your sister? Is she all right?'

'It's fine,' piped up Dave. 'I've put her on the viewing platform with a Coca-Cola and some crisps from the dressing room. She's really happy. Claire and Geronimo are in the production office with the Feastival ops team. Seemed to be enjoying themselves, so I took them another case of Britpop. Ten minutes to show. Shall we get this area clear?'

'Better had.'

'Not a problem.'

'And Dave?'

'Yes, boss?'

'Thanks, man.'

'It's an absolute pleasure,' he said. 'The kids are great.' And his face lit up. I really think he meant it. He put Arte the right way up and, along with the rest of the security detail, ushered him and everyone else out towards front of house.

Showtime.

~

Clearly, the twins were having the best day of their lives. And Claire was with her own gang, and Geronimo was with his people. But I wondered what Beatrix would make of it all. She was only twelve, and all on her own in a strange place. Her older sister had stayed at home, just not interested. I figured Bea would either love it or hate it.

It was her I was thinking about as I walked on the stage. I knew where she was, on the platform next to the sound desk, but I couldn't see her. You can't really see anything when the lights come up. They're too bright. Sometimes you can't hear anything, either. Just a kind of distant rumble, but over the years you learn how to work round it.

Five or six songs in, the whole house was letting go like there was no tomorrow, and I was lost in music when the lights came up on the audience for the first time. I glanced immediately to the platform to see if I could spot Beatrix, and there she was, right at the front, arms aloft, throwing her head back and forward with pure joy and elation painted all over her face and, for the first time in a long time, I cried.

~

A '118' is when you walk straight off the stage, into a vehicle and drive off while the crowd are still shouting for more, beating traffic and temptation in one fell swoop. I high-fived the band, said: 'School night. Gotta do a 118,' and jumped into the driving seat of the family van, which a driver had kindly fetched from the hotel and parked in the loading bay, lights on, engine running. Dave appeared instantly with Claire, the kids and a fresh T-shirt for me. We pulled away from the cheering and rolled out into the stillness of Sunday night.

Arte was in the front. He'd sobered up by then. 'Oh my God, Dad, that was fucking brilliant!'

'Don't swear.'

'Dad, but it was! It was FUCKING BRILLIANT!'

'It was, it was,' everyone in the back agreed. They were all exhilarated and overflowing with excitement and questions.

When would we see Dave again? Soon. Could we get

Over the Rainbow

KFC? Definitely. Could they come to the next show? Maybe. Why didn't you tell us you could do that?

'You hate it when I play the guitar at home. You always tell me to shut up.'

And the chatter went on as far as the M40 when they'd all collapsed into sleep like dogs after a day at the beach.

I looked in the rear-view mirror and smiled.

It might be the best after-show party I've ever been to.

8.

Going Large

June

Contrary to my expectations, the kids were all raring to go in the morning. Uncharacteristically communicative and full of vigour. I had breakfast with them, and was at the weekly farm meeting for 8 a.m. Claire did the school run and then we met in the bath, a bit later than usual, to run through the diary.

Beyond the remaining warm-ups in the coming fortnight, the rest of the shows – the Wembleys and the summer festivals – were almost all at weekends, which at least gave us the weeks in between to concentrate on Feastival. We needed to grab all the time we could to set up another enormous party at home.

First things first: there was work to be done on the time machine. I say time machine – all it was really going to do was take you back about six minutes and keep you there, as long as you were in the Cheese Hub. That was the plan, anyway.

The pipe nightmare in Railway Field was still giving me cold sweats, but the pipe dream Matt Parker and I were hatching in Front Field made all that worthwhile.

The previous year we'd talked ourselves into baubling

Over the Rainbow

out the Cheese Hub with spinning supersymmetric mirror balls: a big success that I was keen to build on.

I'd always wanted to make a massive sundial. Sundials stretch right back into the mists of time. They were mankind's first stab at cosmology. That's what I'm talking about.

Luckily, Matt's wife, Professor Lucie Green, is a leading solar physicist, and it was all coming together quite nicely.

Because it would be built in front of the Cheese Hub, about six minutes west of the Greenwich Meridian, the sundial wouldn't tell the actual time, it would only tell Cheese Hub time. Which, as I say, was about six minutes in the past.

Turns out, it was all dead simple. I went straight up to the Cheese Hub to extrapolate the analemma and then it was right back to giant Frazzles. I got Flava Dave in from the garden and we did a bit of digging around and noticed Frazzles were following me on social media. Result.

I DMed them straight away: 'How about we collaborate on a mega-monster Frazzle-type-thing, a big old party-style whopper, you know, really, really, really huge. It's Cheese Hub Time!'

Frazzles replied immediately. Literally before I was back at my desk with a fresh cuppa.

I clicked on the Frazzles DM.

It said: 'This would not be of interest to us.'

~

Going Large

Sad thing was, a preposterously huge Frazzle would make a lot of practical sense at Feastival.

The Cheese Hub's signature dish is the 'Steakation Sandwich'. It's an evergreen triumph of fine ingredients and simplicity: Blue Monday cheese and aged beef with a bit of Claire's garden salad for colour and texture, in a bun. Boom. A big-hitting flavourbomb that you can hold in one hand while you hold a drink in the other.

Buns work well at festivals because they don't require cutlery. Cutlery is a nightmare at outdoor events. Soon as you start messing with cutlery, you've got to start messing with tables, and then there's no room to dance, so everyone starts dancing on the tables and it's all out of control again, and all because of some stupid bamboo knives and forks that no one likes anyway.

Even plates are a nuisance. We use compostable ones, but they're not really very practical or very sexy and they're certainly not cheap. They're practically the most expensive element of the whole dish and they just get thrown away.

Even if it was a compost bin they were thrown in, it still seemed like a massive waste of energy and there's a thousand festivals up and down the UK alone every summer and everyone is messing about with stupid compostable crockery and cutlery.

Edible plates were clearly the sensible option, and Frazzles seemed the obvious place to start. Beatrix had taken to eating them with our blue cheese on at bedtime, and it had really caught on at home. Soon everyone was saying it's a shame Frazzles aren't just a bit bigger, really like a proper cracker, and then I thought, actually, why not go the whole hog and make them plate-sized so we wouldn't have to muck about with bamboo and they'd soak up all those steakation juices and nothing would get thrown away.

British crisps are the best in the world. You won't find a Monster Munch anywhere else, or even a Hula Hoop, let alone a Frazzle. There are few things that give so much joy for so little money as a bag of classic British crisps, and yet, it is as if we have always been afraid to celebrate them.

There are countless books about wine, about cheese, about chocolate, but I've never, ever seen a book about crisps. And there's nothing on the internet, either, and I think it may be because great crisp recipes are basically trade secrets.

You can trademark a name, but you can't copyright a recipe, so there's nothing to stop anyone making an exact copy of any food whatsoever, from Coca-Cola to KFC, as long as you call it something different.

It's a cast-iron legal requirement to list ingredients on retail food packaging and the discount supermarkets will literally back-engineer their dupes from those ingredient lists and there is nothing to stop them.

I must have burned through a dozen Michelin stars' worth of chefs trying to get to the bottom of how Frazzles are manufactured, and no one could tell me: some mad Willy Wonka machinery at play on yet another monastic miracle, no doubt, but I was getting nowhere.

And now I had to go to Belgium and Holland and Denmark and France. And then Wembley.

~

Going Large

You can't just walk on to the stage at the national stadium. It needs to be a running jump. Warm-up gigs are vital.

Each show we did was basically double the size of the previous one. A tiny church in Colchester, a small theatre in Eastbourne, then a large theatre, an arena, and so on. It's more for the road crew's benefit than for the band. The front of house and stage sound teams need time to make sure everything runs smoothly and, as the shows get bigger, to start to layer on the production elements, cameras and screens and articulated lorries full of delay towers that you need to really rock a big old shed like Wembley.

It was all going well. Really well. Beyond all expectations. There were absolutely no hitches, glitches or squabbles. Just ecstatic audiences, old friends, long-lost cheeses to reconnect with and the glorious sense of freedom that travelling brings.

~

Amsterdam was the first big show: an enormo-dome job with full production. Edd, the sound engineer I'd been working with on Ruby and Violet's songs, lives in Amsterdam. He came to soundcheck bearing a big bag of Dutch specialities. Then Claire and all my friends arrived.

Big stage. Big screens. You can't actually see what's happening on the screens when you're in the band. You're facing the wrong way. So anything could be happening. I was just delighted to be on stage even if I couldn't see what was going on. It's ridiculous how music affects people. Same with my children, same with my friends. Complete exhilaration. Complete release. Completely bonkers.

As I walked off stage, I realized that all I had to do the next day was get to Copenhagen. The next show wasn't until Friday, but there was absolutely no point going home for one night. Apart from that one plane journey, I had two entire days off. No kids. No responsibilities, and I was in Amsterdam, and so was Claire, and so were my friends and it was only midnight, and it was only Tuesday. Complete exhilaration. Complete release.

~

And then it was another day, and I was in Copenhagen in an utterly spectacular hotel, possibly the finest I have ever seen. It seemed to be some kind of converted palace, and there was a huge bunch of flowers and an iced-up bucket of champagne, tower of fruit, box of chocolates and all the rest. It was just tickety-boo, but I had absolutely no idea at all how I'd got there. Or what day it was.

I called Dave. 'This is Copenhagen, right?'

'Yeah. You all right? Damon was a bit worried about you.'

'Oh, fuck!'

'Maybe write him a note.'

'OK. So. Fuck. Right. I remember that bar with all the margaritas and then I remember walking back to a completely different hotel and it was miles and then it started raining, right? And we bought some umbrellas? We were singing in the rain, right? But that was good?'

'It was quite funny.'

'And we were in bed by like six, I remember, and then I definitely got a solid couple of hours because I took one of those sleeping tablets Edd gave me.'

'Yeah, you were pretty sleepy. Fine. Bit sleepy, though.'

'Dave?'

'Yes, Alex.'

'Thanks, man.'

'You're welcome. As I said. It really is a pleasure.' I couldn't see his face, but I really think he meant it.

~

I had a sauna and a steam, then went to the nearest massive supermarket in a taxi. I was having a hotel picnic on the bed when Claire arrived, exhausted, with the kids. They were all beyond excited for Bea's birthday, and for Roskilde.

Roskilde is one of the world's great festivals. Brilliant line-up, brilliantly run and absolutely huge. All the kids' favourite bands seemed to be playing, but the cherry on Bea's birthday cake was that their godfather – their favourite person, and mine, Robert – was coming with his family.

We met them all at Tivoli Gardens, the world's greatest theme park, the one that Disneyland was based on. It's got a Michelin-starred restaurant at one end and the world's best hot dogs at the other. It's got all kinds of swirling, twirling, spinning machines, and Hans Christian Andersen's ghost, and liquorice, and lasers, and fountains and flowers. It's a family festival that never stops. We stayed all day and then we stayed all evening and then there was fireworks and then we all jumped in the river, and we all went to bed.

The next day, I left early for soundcheck and told everyone I'd see them after the show was over. The kids had been instructed to leave no stone unturned, explore every corner and see if there were any good things happening at Roskilde that we weren't already doing at Feastival.

I spoke to the news cameras and told them how delighted I was to be back in Denmark, and I spoke to Damon and told him how delighted I was to be in Denmark, all things considered, and he said, 'Hmmm,' and then we played one of the best shows we have ever played.

The crowd was vast. They were singing before we started. The noise was deafening. Massed elation and tears and laughter and letting go, which can't fail to heal.

About halfway through, Damon led the entire ecstatic crowd in a rendition of 'Happy Birthday' to Beatrix, which I figured meant I was off the hook.

I was delighted to see all my lot were in really good shape when they came to the dressing room with Robert afterwards. Bea was incandescent with joy, but Robert was looking slightly worried.

'Where are your lot?' I said. He's got twin girls a year older than Arte and Gali, and a son a year younger.

He said they were all stuck in a tree in the VIP glamping area.

Dave, who had been rechristened 'Saver' Dave since marshalling me safely to this land of Lego and plenty was scrambled and led the rescue party.

Going Large

They were gone for a long time. Eventually, Dave called to say he'd got them all out of the tree fine, but now they'd found the rave stage and were doing squares and enjoying themselves, and yes, he'd look after them and it was a pleasure. 'They're great kids. That was some tree they all climbed, by the way.'

I could tell he was chuckling to himself.

~

We took the train out to the family home of Mette, Robert's wife, for late lunch the next day. Her dad was there, and her sister, and Robert cooked. The kids tore off on a mass of bicycles and we walked down to the harbour to meet them. We jumped in the sea and walked home in the moonlight for hot chocolates and peace and plans and news and dreams and that honestly felt better than playing to all those people had the night before.

I needed that anchor of friends and family more than ever that night. It had been a huge week. Roskilde had been a triumph – a front-page newspaper headline that morning had announced 'Blur – the best band in the world!' – but Damon wasn't at all happy with the way the show had gone in Amsterdam and had left immediately afterwards: an unscheduled 118. We'd had to start two songs again because they'd gone wrong halfway through. That's not even really OK in a warm-up show, let alone in Holland's biggest arena. Fair to say none of my guests had seemed to notice, they'd loved the whole thing. But audience reaction wasn't the reason the band hadn't been performing for all these years.

Over the Rainbow

That was basically because mainly Damon – and, occasionally, Graham – hadn't wanted to. If we couldn't play literally better than we ever had before – Damon, particularly, isn't interested in things that aren't moving forward – then that would be it, forever. I had to go all in. All the music and all the joy and laughter this year had brought had made it very clear how much there was to play for.

9.
Wembley

July

It wasn't until after I'd pressed send on the text to the tour manager saying: 'We're on our way to Wembley,' that I realized what it said – that this was a pretty special moment. I started singing.

We were *on our way to Wembley*, the whole family.

The band getting back together had cast a spell that was working across all aspects of my life. I was fitter. I was feeling good and so were Claire and the kids. It had brought us all closer together.

Obviously they were all telling me to stop singing and shut up, but they were full of excitement, brimming with glee.

Claire and I, even though we had been running in different directions and still had to conduct our morning meetings in the bath after the gym, were, I think, closer than ever.

While I was running around, she was running the farm and the festival and the family, and that made me happier than anything, other than the fact the kids were happy.

They were having the summer of their lives, and we were on our way to Wembley.

Over the Rainbow

~

We headed down to London a day early because, due to some miscalculation in his tennis game, 'Labour' Dave Rowntree had kippered his knee and couldn't walk. Although he could still play drums, just.

We'd had to cancel the show immediately prior to Wembley, a festival in France, on his doctor's orders, so the entire production had arrived at the stadium for soundcheck a day earlier than expected, and we had the whole place to ourselves. Like the film *Night at the Museum*, if it had been set in a stadium, in the daytime.

I don't think any Englishman can approach the fabled arch without an involuntary visceral response. You feel it in your stomach because Wembley is emblematic. It was designed to be. It's daunting. It's vast. This is a venue purpose-built to stage events of national importance. By the time the kids were singing 'We're on our way to Wembley' I'd gone a little bit quiet.

We were waved through several security cordons, greeted by our own local security manager and transferred to buggies that zoomed along endless corridors, through shoaling high-vis jackets, pink, yellow, grey and green; past catering areas, through crew rest zones, production offices, hospitality suites for VIPs; on and on, through security checkpoint after security checkpoint, past the support bands' dressing rooms, finally arriving with a jolt at the home team's dressing room. The England team's dressing room.

I was the first one to arrive, as usual.

Lydia, who looks after the band's dressing room, was

Wembley

waiting at the door with a hot cup of tea and a freshly cracked pack of Camels with one poking out. 'Just so you know, they're a bit funny about smoking here,' she said, as she fiddled with a lighter and took my bag.

The kids had dashed past her and there were whoops of excitement coming from beyond the door.

'You need to be OUT of here before the band arrives,' I reminded them. 'You know the rules. Don't make a mess. Put that back, Arte, and SHUT the fridge. It's eleven o'clock in the fucking morning.'

There were bunches of flowers from record companies, from publishers, from agents. There were stacks of parcels from fashion houses, from audio companies and footballs and England shirts with our surnames on the back. Fresh pants, fresh socks, toothbrushes.

It was like Christmas morning in there, and Boxing Day as well, because there was the full Boxing Day buffet laid out. And there were cakes, crisps, sweets, chocolate bars, chocolate boxes, a juice station, a two-group commercial espresso machine, fridges full of fizzy pop, display fridges full of beer, branded fridges full of wine, a freezer full of preposterously expensive vodka, and a cocktail bar.

'Do you want anything to eat?' said Lydia. 'Band catering is just next door.'

'Are you kidding?' I said. 'I've got an appointment with my Britpop trousers tomorrow. What are we going to do if they don't fit?'

'Drink, then?'

'Still on a yellow, from Amsterdam.'

'Well, I've got the Britpop on ice, anyway.'

'Thanks, Lyds.'

Over the Rainbow

'Pleasure. Ooh, you'll like this. Look!' She led me through another door into what was, basically, a huge spa. Massage tables, treatment rooms and mammoth communal shower for a dozen muddy footballers. And, best of all, hot tubs and ice baths.

'I'm not going to have a drink till Sunday night. 118 buggy direct to this very room tomorrow, right? Showers all on. Hot tubs. Cold tubs, all bubbling. I'm going to need the full works.'

'Copy that, and Richie's ready on stage when you are. Beatrix has your buggy keys by the way.'

'Oh, Jesus!'

Right up until the moment they discover drinking, I think kids like buggies more than anything else. I suspect no child is immune to buggy fever. They love them and they just want to drive them as fast as they can.

We get two allocated at Feastival, one for me and one for Claire, and we have to wear the keys around our necks because if the slightest opportunity presents itself there is instantly a buggy haring around at high speed overloaded with as many delighted children as can possibly sardine into it, hang off the back of it and sit on the roof.

Dave was scrambled on the radio to requisition another buggy and go in search of Beatrix, who was hopefully still somewhere within the stadium, and I took a stroll to the stage.

I figured it would be nice to walk and soak it all in, but it wasn't until I began to tackle Wembley as a pedestrian that its galactic scale really hit home.

It's monumental. It is actually a monument.

I hadn't even got out of backstage yet, but as I walked

Wembley

up the ramp I caught another glimpse of the arch, the 360 wraparound triple tiers of grandstand, the lighting rig, screens as big as football pitches and the pitch itself, covered over and tiny in the distance.

It suddenly felt like a tall order for two guitars, a keyboard and a drummer on crutches, and my knees were going all trembly

And then two things happened. Whoever was line-checking the keyboard started playing a football song, 'Ozzie's Going To Wembley', and then, when they'd finished, I picked up my bass and I played 'Vindaloo' and it felt good.

I got goosebumps of a completely different kind. Wembley is the spiritual home of that song. They've been singing it here for years, and I suddenly felt like I'd earned the right to walk on to that stage and let it rip.

Just as I realized I wasn't in the least bit worried, Beatrix came careering out of the tunnel and started doing doughnuts somewhere around where the centre circle must have been hidden.

I legged it down the ramp, vaulted the crowd barrier and tore up the pitch to catch her and handed her over to Saver Dave who had evidently been in hot pursuit for some time.

'Big, innit?!' he said, and as I peered at the business end, three tiny figures, my bandmates, appeared on the stage in the distance.

I jogged back to join them and we got stuck right in. Just ran through the entire set to an empty stadium.

Empty apart from the production team, Claire, the kids, and Ruby and Violet and Bianca. They didn't stop bouncing around, running around, flailing and moshing.

Over the Rainbow

By the time we'd run through the set the venue didn't feel anywhere near as big as it had. It felt very comfortable on stage. Backstage, though, guests and guest lists were proving an absolute nightmare, as usual. Guest lists are the single trickiest element of the entire situation. Even more tedious than drains. Dealing with guest passes, guest access, guest parking and guests' requests was a non-stop fiddlefest of calls, texts, emails and WhatsApps, which Georgie from the office and Geronimo were slogging through.

From time to time, they'd show me the more diva-ish demands.

'Will we be standing with everyone else?'

'And we'd all love to come for lunch at the farm the following day if possible?'

'How do we get home afterwards?'

Weeks of endless stupid questions, and now Georgie and Geronimo were going for a dedicated briefing with all the other poor souls who were dealing with all the other guest lists.

The rest of the family piled back to the Wembley Hilton, where we were staying for the weekend. It is the world's greyest hotel, but it was closer to the dressing room than the stage.

~

Given that the farm is about as isolated as you could hope to be within a hundred miles of the biggest city in Europe, being in any kind of inner-city environment, particularly on a Friday night in the summertime, was quite an

Wembley

exciting prospect for the whole family. The Eliots, who live in their own splendid isolation in Cornwall, had done some Googling and were cock-a-hoop that there was a Costco within walking distance. They insisted we get the party started there.

I've never been to a Costco, but even before we'd got inside, as soon as I saw how big the trolleys were, I knew I was going to like it.

It was to all other supermarkets what Wembley is to all other concert venues.

Massive tellies, walk-in fridges, inflatable hot tubs, canoes: near the tills where they normally put the impulse buys there was a forest of two-hundred-year-old olive trees. Literally, everything we needed and all at really tempting prices. All the food was sold by the crate, the barrel and the dozen.

We were more excited in there than any of us had been in Wembley. Claire and I went back to the hotel with as much as stuff as we could fit in the van, and the kids disappeared into the night.

Claire put the Britpop trousers on the end of the bed, like a stocking on Christmas Eve, and the room was full of Costco goodies, and I wondered how I'd ever get to sleep. But then the alarm was going off and it was 8 o'clock in the morning and it was time to go.

~

The Britpop trousers. They actually fitted. I moonwalked up Wembley Way with Saver Dave. As we entered the venue, one of the support acts was sound-checking.

Over the Rainbow

The first thing I noticed was that it was incredibly loud. We hadn't been able to crank the sound system the day before because of noise restrictions, but as this was a show day, those were all lifted and the band – whoever they were – were going full throttle.

Second thing was that it sounded absolutely incredible. So incredible that I was suddenly filled with doubt. This band were just grooving, but playing with such swagger and confidence it was staggering. Like they owned the place. Like they were the house band.

'Who the hell is this and how the hell are we ever going to compete with them?!' I shouted to Dave.

He got straight on the radio to production to find out, and then held his radio up to his ear so he could make out their response. He turned away, but I could see he was laughing.

'It's a recording of you lot, mucking around in soundcheck yesterday.'

We had until doors opened to run through 'Tender' with the London Community Gospel Choir. There was about a hundred of them, and it took less time to rehearse the song than it did to work out how to get them all on and off stage calmly and swiftly – more like practising a fire drill than rehearsing a big moment.

It was, in fact, all very enjoyable, and had been since we'd arrived at the venue the day before. The band were on good form, hitting our stride and making each other laugh again. It didn't feel like work at all. It felt more like I was on holiday than actually being on holiday does. The kids were all happy, and I didn't have to drive anyone to McDonald's or do any washing-up at all.

Wembley

Better still, with soundcheck done, a rare and wonderful opportunity presented itself: an afternoon nap in the dressing room.

~

As I slept, pubs and bars around Wembley filled with Blur fans and then, slowly, decanted them into the stadium. They trickled in from across the city, the country, the world; thousands and thousands of people, wonderful people, young and old, rich and poor, who all seemed to care about this thing as much as we all did.

I woke up mid-afternoon. The first support band was on, and guests had started to arrive. I put all the showers on and ran around singing Roy Orbison's 'Dream Baby' – the reverb was fantastic in there. Then I had a hot tub and a cold tub and a haircut and all of a sudden it was time to go.

We were ready and London was waiting.

~

It's hard to recall without getting goosepimples. London had clearly been anticipating this moment for as long as we had. Everyone who was there was there to have a good time, and the band were equal to it.

Damon was completely at home in front of that crowd of nearly 100,000 people. Totally comfortable in his skin. Absolutely confident in the music. I could feel his joy. It lit up the whole stadium: his exhilaration was one-hundred-per-cent infectious. No one who could see or hear him was immune.

Over the Rainbow

Graham, the best guitar player I've ever heard, was just ringing a bell, as calm and relaxed as he would have been if there had been no one watching at all.

And Rowntree, his crutches leaning on the riser, was battering the living daylights out of the kit.

There was one moment early on, just as darkness was gathering over north-west London. We were playing 'Out of Time', a song I'd really pushed to be included in the set. I was playing with my eyes closed, really concentrating, and I thought it had gone a bit quiet.

I opened my eyes, and lights, tens of thousands of lights, were shining back at me from the grandstands. The whole place was electrified.

It all made sense. That huge amphitheatre, all those tiers of seating, the stage and the screens; Wembley resonating at a brilliant hum that built and built to crescendo finale.

We played 'The Narcissist', the new release from *The Ballad of Darren*. It really landed. The audience sang every single word, I was still reeling from that when suddenly, somehow, it was the last song already: 'The Universal'. And in the two-beat pause before the instrumental outro, a super trouper lit up a strange shape in the crowd. It was right at the front, and right in my eyeline, and I realized it was Arte on his twin brother's shoulders, his spotlit face painted with pure elation.

~

As a convoy of buggies hurtled back to the dressing room at unnecessary speed, horns blaring, the entire stadium

Wembley

ringing, ecstatic, the boss – the crowned king of Wembley Stadium – was by my side, eyes brilliant, a massive grin all over his face.

'We've got to have a drink tonight, right?' he said.

It would have been rude not to.

I had another quick run around the showers and a Britpop or two in the dressing room, then nipped up to hospitality to say hello to as many people as possible, but it was total mayhem. All I wanted, in that moment, was peace and quiet.

I snuck back downstairs to get another bottle of Britpop and the only person in the dressing room was Banksy, and that was much nicer, and I was all tucked up in bed by four. Ish.

~

There was a soundcheck on Sunday morning. We planned to run through a few songs we hadn't tried for a while, to mix it up a bit for the second show. When the alarm went off at nine, my mental state was an interesting mix of Britpop-induced wretchedness and Wembley triumph. But playing music is the best way to assuage a hangover.

By the time we'd trotted through a few songs, had another run around the showers, a hearty breakfast and read the reviews, I was just about ready for another nap but ended up watching *Chitty Chitty Bang Bang* with Graham, which was even better.

10.

High Summer

July, August

We were all back at the farm by lunchtime the following day, for the eve of school holidays and the start of the festival season proper. The final Feastival PR push meant I spent most of the following week doing back-to-back 'phoners' – interviews over the telephone – 'Zoomers', and radio and television appearances. Then there was a family photo shoot at the farm on the Thursday – it's a family business, after all – that had to be dealt with in record time as I had to leave mid-afternoon for a Blur show at a festival in deepest rural France, miles from anywhere. I was completely exhausted.

~

The gig was only a couple of hundred miles from home as the crow flies, but there was no practical way to get there and back quickly. It was a smidge too far for a chopper, and too remote from a twenty-four-hour aerodrome for a charter to be any quicker. So, we took a sleeper bus. A big old, smelly old tour bus, which is like living on a boat with the people that you work with. It's like living on a submarine with your whole family. It's like both of those things at once. It's horrible.

It was raining hard when we arrived on site at dawn, but it was Bastille Day, the big French summer holiday, and they were keen to make it count.

I could have stayed there all summer. 'Rural France in the rain' is my idea of a perfect holiday. But it was just a tantalizing, fleeting kiss from La Belle France: we played at midnight in warm wind and rain and then did a 118.

I knew tour buses were horrible. But I'd forgotten how horrible horrible can actually be. Maybe everyone else had, too.

Because there was a mutiny. The tour manager was overruled and the driver directed straight to Paris so that we could get the Eurostar to London, rather than schlepping all around the Dover Strait again.

We pulled in opposite Gare du Nord at the crack of dawn and I managed to grab a suitcase full of baguettes and patisserie before we all jumped on the first train to St Pancras.

~

I made a connecting train back to the farm with seconds to spare, and when I got back to the farm it was still quite early. I couldn't raise Claire or any of the kids to come and fetch me on the quad bike, so I lugged my suitcase and half a boulangerie over Railway Field, noticing with satisfaction that the drainage works were complete.

I woke the kids up, put a ham on to simmer with some peppercorns and a handful of bay leaves and thyme from the garden, did a truly epic amount of washing-up, and then woke the kids up again.

I sliced the finished ham on the circular slicer and piled it all up in the main fridge, stacked the French goodies on the butcher's block in the middle of the kitchen, and then collapsed in bed until my alarm went off at 5.00 a.m. the following day.

The house was still and silent, and I smoked a perfect cigarette in the glow of dawn. It was so good to be home.

All the ham and all the bread had gone and there was another huge stack of washing-up sitting in that same warm sunshine, like a perfect, postmodern still life.

~

Call time for *Sunday Brunch* was 8.00 a.m. at Television Centre, literally a stone's throw from Studio Thirteen.

As the car passed Damon's studio and turned right onto Wood Lane, it seemed a very long time since the exhilaration and the promise of that first album playback just before Easter. It suddenly occurred to me, with Wembley done, we were nearer the end of the road than the beginning.

It had been a gift, spending time with the band, totally focused on something that I really cared about; it had somehow brought not only my oldest, dearest friends – the band – closer again, but my family, too.

I'd been so consumed by what was happening that it had been hard to see beyond it. I really hadn't thought about how it would all end, or what would happen after the record was out and the gigs were done.

Over the Rainbow

The pressure was on. We'd won the battle of Wembley, but still had the album campaign to negotiate, and travel fatigue was starting to kick in.

Everyone seemed to be bearing up OK, but then I'm quite self-absorbed at the best of times.

From my experience of being one of four blokes in a band, when any one of them is unhappy, rather than speak out and share their frustrations they tend to go quieter and quieter and then blow an almighty gasket.

We'd never got to the end of a tour or album cycle without at least one of the band declaring they'd had enough.

But it was such a special thing. A fact I couldn't have forgotten, even if I'd wanted to.

~

One of the other guests on *Sunday Brunch* had been at Wembley and got quite emotional when we were introduced at the top of the programme. She kept going on about how brilliant the show had been, but I was really there to talk about my Cheese Hub Time Machine and how brilliant that was going to be, and it was difficult trying to stay on message.

I was trying to explain how local apparent noon is calculated and the difference between Greenwich Mean Time and Cheese Hub Good Time, but my phone kept buzzing in my pocket. Fortunately it was on silent, but it was clearly something very important. It was whirring away like an electric toothbrush. As soon as we went to the break, I dashed off set in a panic to see what the emergency was.

High Summer

Beatrix had messaged the 'Jameses on Tour' family WhatsApp group and put in a request for me to pick up a KFC from Peartree Roundabout on the way home, and there'd been a big pile-on of specific requests for extra hot wings, extra popcorn chicken, could I maybe pop into Greggs as well and finally one from Claire saying let us know what time you're on and we'll all try to watch it.

I agreed to all their demands but only if all the washing-up was done by the time I got back.

~

We had a big old KFC picnic in the cherry orchard. Those cherries were absolutely delicious with the KFC, in the Coca-Cola, in the Britpop. They were, it seemed, endlessly versatile and there were a hell of a lot of them. Way, way more than I'd ever seen before.

Claire said that the whole market garden was like that this year. It had all suddenly gone haywire. Everything was super-cropping and it must be because the bees were finally doing the business.

We went for a walk around the garden and, indeed, it was a completely different garden from the one I remembered.

The bee project had been a long, long labour of love, masterminded by Claire and we were five summers and endless bills in, and I still hadn't seen a speck of honey, but Claire had refused to give up. It seemed that, finally, the hard work was paying off.

Not only had she spun a quarter of a tonne of honey while I'd been in France, but yields across the entire market garden had gone completely through the roof from the extra pollinating the bees were doing.

Honey is the nectar of the gods, the spirit of the garden, distilled: miraculous stuff, Claire was saying. Not just to me. To everybody, all the time.

True: it is not just delicious; it has all kinds of health benefits and natural antigens. It's good with cheese. It doesn't have a shelf life, either. They found it buried with the Pharaohs and it was still edible. Cleopatra used it as a face mask. There was plenty of it, so we all tried that. Pretty good.

Fair to say, the more I found out about honey, the more I liked it.

But what was I going to do with a quarter of a tonne of the stuff, which, as the weeks went by rose to nearer half a tonne? Storage was becoming a problem. Bees are very good at finding honey and telling all the other bees where it is.

And what was I going to do with half a tonne of cherries?

But that's always the trouble with farming. If something doesn't work, it's a problem, obviously, but it's when it does work that your problems really start.

I needed to sleep on that little problem.

I needed to sleep for about a hundred years, all of a sudden.

The kids were on holiday, the house was mayhem, and Blur were playing festivals every weekend until Feastival, which was suddenly approaching at breakneck speed.

All hands on deck. We put the kids to work in the garden or the office or on Feastival and Cheese Hub planning

High Summer

during the week, and then just moved the whole operation to wherever Blur were playing each weekend – family, office, the whole business.

It actually worked pretty well.

~

We set off on our first festival of the holidays. The whole family and Georgie from the office flying into Lucca, Italy, a day early.

The promoter had kindly booked us the ambassadorial suite in a palazzo right on the main square. The square was all fenced off because there was a concert going on there later, but there was a Michelin-starred restaurant right next door to the hotel. We were all famished so we headed straight there, sat on the terrace and ordered the tasting menu with all the wine pairings for all the family if that was OK and the waiter said yes, of course, it was perfect.

It was indeed, already, exactly perfect when a band struck up beyond the hoardings blocking off the square.

'OHMYGODITSLILNASX,' said Geronimo, jumping to his feet. 'We've GOT to go.'

Lil Nas X was their favourite singer. 'Old Town Road' was right up there with 'Rasputin' in the all-time family favourites.

I called the new tour manager and asked if it was too late to buy tickets, and he called me back two minutes later, and said it was the same promoter, and he'd put all the kids' names on the door nearest the restaurant, and they thought that was pretty cool, and left their dessert wines and their tiramisus on the table and disappeared.

Over the Rainbow

I ordered espressos and cognacs and a cigar and the waiter said, 'Lot of children! All yours?' and he raised his eyebrows and left the bottle of cognac on the table with a wink. He was asking us if we wanted anything else, and we were discussing the cheese options, when 'Old Town Road' kicked in and everyone – waiting staff, customers, the entire brigade from the kitchen – came out to sing along. I did, too, basking in how nice it was to be part of a singing crowd on the other end of a banger, live, as it happened, in the summertime.

~

It was a late show the next day. Temperatures in the daytime had hit forty degrees and it was still roasting when we went on stage after midnight. It was by far the biggest show we've ever done in Italy and the most rapturously received.

A *Vanity Fair* party's worth of headline-grabbing A-listers had appeared as if by magic out of the Tuscan hillsides to see the band, and by about 4.00 a.m. the party in the ambassadorial suite was peaking. Geronimo was DJing. Claire had rallied round the night staff, who seemed more than happy to help and had appeared in a jiffy with a wheeled gondola loaded with limoncello and grappa on a big bed of ice.

The rest of the band were getting the early flight, but the Jameses stuck around for a late breakfast and headed out to the airport at lunchtime, all feeling pretty shattered.

High Summer

~

The rest of the family were going home, but my destination was a hotel in Soho. I had to do Chris Moyles's breakfast show the following morning in Leicester Square, to talk about Feastival, and then there was the live performance of *The Ballad of Darren* in its entirety at the Hammersmith Apollo that evening.

It had just come out and it was number one all over the place.

We had a rehearsal in the afternoon, the first full runthrough. It was a big day and potentially a tricky one, so I'd planned to go straight to bed when I got to Soho – which should have been about 6.00 p.m., but there was a problem with the plane, and we were still sitting in the airport at midnight. Michelin stars and groaning gondolas were a dim and distant memory. There was just a vending machine that had run out of everything except Pringles.

It's impossible to sleep, or even relax, at an airport when your plane is delayed and you've got five teenagers and Pringles for company. We landed at Gatwick at 2.00 a.m. I kissed Claire goodbye and took Geronimo to Soho with me to make sure I got up in time. We got to the hotel by four. Geronimo went to Bar Italia and drank espressos until six when he woke me up with a cappuccino, as planned. I must have got a solid – vital – ninety minutes in.

I beetled through Soho and Chinatown's sweet August stickiness and got there bang on time. The doorman at Global Radio high-fived me. I used to work there.

'Bit early for you ain't it?'

I nodded. Bleary.

Every time you walk through the doors of a broadcast facility you have to remind yourself you are entering what is effectively a live minefield. It can be precarious doing promo mid-tour. If you say the wrong thing, it can haunt you forever.

I remember doing the biggest breakfast show on German radio with Damon back in the nineties after a great gig and a big night in Berlin. Damon fell sound asleep mid-sentence live on air to millions of breakfasting Germans.

And then, on what was maybe the second or third year of Feastival, I'd whizzed down to Broadcasting House right in the middle of the build to appear on Jo Whiley's BBC Radio show. She introduced me with great fanfare and asked, 'So, Alex, tell us what we can expect from Feastival?' and I said nothing. Nothing at all. For about ten whole seconds. I'm not totally sure, but I think broadcasters get fined more for broadcasting silences than for broadcasting swearing. It's certainly a bigger crime.

So, with Herculean effort I'd focused my thoughts, concentrated with all my might and said, 'Meh . . .'

I was literally so tired I couldn't speak. I was all done. Exhausted. Spent.

And that was how I felt now.

But Moylesey completely put me at my ease. Within about ten seconds I'd completely forgotten I'd had ninety minutes' sleep and nothing to eat except a sour-cream-flavour Pringle about twelve hours ago, and we just rattled on like no one was listening and I couldn't stop laughing or telling the truth or remembering another funny thing

High Summer

that had happened and somehow or other it was all OK. We ran over the allocated time slot and I just made it in time for the band call at the Hammersmith Apollo.

This gig – the album gig – was the last big Blur hurdle. Ideally, I'd have had more than ninety minutes' sleep.

We'd been rehearsing the songs from *The Ballad of Darren* in soundchecks. We'd played two of them as part of the set at Wembley, but only two. We'd chucked one or two more in here and there but we hadn't even *tried* to play the whole thing through from beginning to end. It can be tricky to realize studio material. On the previous album we'd spent longer in the rehearsal studio, trying to figure out how to play it live, than we did in the recording studio actually making it.

We still hadn't quite managed to run through the whole record without stopping by the end of soundcheck, which went right up until doors. But it was feeling pretty good. I grabbed another solid thirty minutes' kip and then, when the moment came, it was like walking out on stage at Wembley, only more intense – the clamouring of the audience hit less in the chest, more in the face. The crowd were baying for our first new body of work for forever and a day. Some of them were crying. And so, I think, were some of the band.

~

We played the entire thing note-perfect. We absolutely smashed it. Which was cause for great celebration. James, the producer, was possibly the most keen of all to celebrate, and when they closed the bar at the Apollo we went to Chiltern Firehouse to do just that.

Over the Rainbow

When we got there I suddenly felt very tired. I left him in the upstairs bar with Claire to find a quiet corner down on the smoking terrace.

I had a margarita and I had a Marlboro and I basked alone in sweet, satisfied exhaustion. On a sticky wicket, the night had been an absolute triumph. In some ways, the biggest triumph of a year of unexpected triumphs.

Sweet, sweet exhaustion.

A voice said, 'Come and sit with us . . .'

Two exceptionally pretty women, even by Firehouse standards, were beckoning me to join them. I just didn't have the energy or the inclination. I was spent, again. I just smiled and nodded and blew out a big old billow of smoke.

'Don't be shy,' said the other one as they moved to join me.

'What's your name?' said the first one and I told them and they asked me what I did and I explained that I was a musician, at least I was a bass player, and yes, I'd literally just done a gig in West London with my band. Didn't say where. Didn't say which.

I couldn't really say what I was thinking out loud because it was something like, 'And I honestly think we might be the greatest band in the world and we might seriously just have done our greatest ever gig. And we didn't really rehearse. And I only got ninety minutes' in. Honestly, it's like magic.'

So, I didn't say anything at all, and it went quiet for a beat or two.

The second one said, 'So, what *else* do you do?'

I said, 'Oh, I make cheese,' and laughed.

'Oh my God, really? We love cheese.'

High Summer

We talked about cheese for ages. But although I could hear myself talking about cheese, I was really just thinking about how much I love my band, and it sank in that, as of that very moment, Blur time was officially running out.

Wembley was done, the album was released, and all of a sudden, it felt like there was no more road. We'd made a great record, not just against the odds: statistically it was nothing short of a miracle. Without a doubt we'd just played the best run of shows of our lives. How we would ever top that in future? I had absolutely no idea, and Damon is only ever interested in going one better.

There were still shows to come in Japan and South America, and one or two European festivals. If we got it all exactly right, the worry was that the feeling would be, 'Well, that would be a good place to leave it.' But there was still plenty of time to mess it up, too. I drifted back up to the bar.

11.

All Back to Mine

August

Normally, the whole of August is all about Feastival. For the past four or five years, I hadn't left the farm all month long. But this year, it was festivals all weekend, every weekend as summer hurtled by. Blur literally never played the same set twice. The back-to-basics approach meant we could change what we were playing at a moment's notice. So, we ran through songs from *The Ballad* and from the dim and distant past in soundchecks and, if they felt good, we incorporated them into the set. The band was a well-oiled machine at this point, but playing at a festival is one thing. Hosting one is another thing altogether.

~

It takes more than a year to plan Feastival, so by the beginning of August we were already well into trying to secure headliners and thinking about preparatory works and planting in the big wood for the following summer. The physical build, though, happens very fast. It only begins in earnest the weekend before the bank holiday.

I've always loved the build: it's dazzling. We're basically

assembling a town with a population of 25,000. That's about the same size as Witney, the largest market town in the Cotswolds. It needs everything that a large market town needs: roads, parking, power, housing, sanitation. It even has its own hospital and its own Marks & Spencer, and that's before you start fiddling about with the stuff that makes it fun: the stages, backstages, lighting and sound rigs, Michelin-starred food, sundials, ballerinas and bumper cars. And the whole thing goes up in a week.

It's exciting to watch over more than a year's careful planning and preparation explode into life.

But this year I'd be watching over it from about 6,000 miles away.

~

When I left the farm for Heathrow, it was raining hard and there was still just one guy with a GPS and a lot of coloured flags in Front Field marking out the positions of everything. He'd been there a few days.

I brought Arte with me. Claire had sent him along partly, I think, to keep an eye on me. And partly because he was family 'employee of the month'. And partly because he's always, always wanted to go to Japan.

I switched my phone on as we touched down nine time zones away at Tokyo Haneda. It started to ring immediately.

It was Claire. The build was in full swing. The entire crew, 3,000 strong, was going at it hammer and tongs. It hadn't stopped raining, and there was an articulated lorry stuck in the middle of the main arena. But, she said, she was just in the Cheese Hub with some mathematicians from Cambridge, Britain's leading solar physicist and a team of builders, and did I want the pink flags or the blue ones to mark off the hours on the sundial, just checking because she had to order them right now.

And then I heard her say, 'ER, NO! STOP THAT! NOW!! YES!!! YOU!!!!! Darling, I'll have to call you back. I love you.'

'You decide what colour. I trust you,' I said and told her I loved her, too, before realizing she'd hung up already.

But I guess she knew that I loved her and that I trusted her and Geronimo to do the right thing. Maybe it's inevitable that you eventually give the things you love the most to the people that you love the most.

~

I wrestled my attention back to the here and now: Japan. It's a country that has always dazzled me: it's got the best spas and the best food on the planet.

Japanese fans are uniquely brilliant, too. I don't know how they manage to find out what flight we are on, what bullet train we are getting, or which hotel we are staying at, but they always have and they always do, and manoeuvring through hotel lobbies, airport terminals and railway stations can quickly descend into a comic caper with a bubbling sea of vinyl-bearing, camera-waving, gift-thrusting devotees, respectful and polite even when absolutely losing it, unable to contain their emotion.

It's hard not to stop and engage a little bit, but things can get out of control quickly. As soon as the first person

Over the Rainbow

in the Blur entourage stops, the entire band party quickly becomes surrounded, and that spectacle draws a crowd of its own at which point it becomes hard to break through the lines and it's a full-on siege.

I could see Saver Dave was relieved to get Arte and me into our van at the airport, but by the time everyone else was in their vans kerbside, the entire stationary cavalcade was mobbed and fans were happily pressing their noses up to the windows and waving.

Dave said, 'Welcome to Japan, Arte.'

~

The hotel was colossal. A glass elevator zoomed us to the silence and serenity of the ninety-ninth floor, just in time for breakfast. I would have felt very isolated if Arte hadn't been with me. He was awestruck by the accommodation and his joy was mine. Where the double aspect floor-to-ceiling windows met in the corner there was a huge sculpture, which on closer inspection turned out to be made entirely from chocolate.

Now Arte was marvelling at the bowl of fruit. Apples, big as Christmas puddings and bright as Christmas baubles, foot-long straight bananas and a peach that would have fed a whole family.

We skipped breakfast and went straight down to the sauna. It was like a furnace and it had a huge television in it. It was good that there was something to pretend to look at because Arte was slightly cowed by the strictly enforced total nakedness policy, which is standard practice in Japanese spas.

All Back to Mine

Afterwards, we met Saver Dave at the Club Lounge in the penthouse. We were all keen to go in search of noodles.

I'd got Arte hooked on farmhouse ramen back in the Britpop-trouser-squeaking-bone-broth days of winter, but Japanese noodles are never quite the same outside of Japan. I think anyone who has acquired a taste for the real thing is doomed to spend the rest of their life yearning for authentic ramen, soba and udon.

The weather was a physical assault. A left–right of intense heat and humidity, combined with the roundhouse of long-haul eastbound jetlag, meant we could hardly walk straight when we got outside via a 117. People and traffic were thronging in all directions. Even the pavements seemed to have slow, fast and overtaking lanes for pedestrians. Fortunately, though, there is a noodle bar on every corner in Shibuya and, sure enough, just down the block we caught the unmistakable, heavenly aroma: the savoury embrace of dashi. The smell alone was as good as a rest, and as we ducked beneath the lanterns, through the tiny doorway, and took our seats at the counter, all the weight of the world fell away.

The word 'restaurant' comes from eighteenth-century France, when the very first establishments opened, selling just one dish: a restorative broth, a *bouillon restaurant*. I grew up on French food, but if I had to pick a favourite of all the world's cuisines I would pick Japanese, no hesitation: the Japanese are the Zen masters of the bouillon restaurant.

There's a gazillion tiny noodle bars in every big city in Japan, and they all have their own unique secret stock recipe: their own variations on the theme. You order from

Over the Rainbow

a machine, which hopefully has pictures on it, and immediately a glass of ice-cold water appears – they really know how to serve water in Japan. Often, you are given a huge bib with the water because it can get quite messy, and within about three minutes of ordering, a steaming bowl as big your head appears, all kinds of delights bobbing on its surface and hidden in its depths, and with an involuntary smile you join the chorus of happy slurpers.

Boy, how we slurped. How we laughed and how we slurped some more. By the time we'd got to the bottom of the bowl we were fully restored.

Arte wanted to go to a cat cafe – and according to Google, we were very close to one.

Unfortunately addresses in Japan make no sense whatsoever, and after walking round in pleasant circles for an hour we found the cat cafe we'd been looking for all along on the fifth floor of the building we'd been in to start with.

For about five quid they give you a plastic cup of coffee and handful of cat lollies and cat nibbles. Then you pass through an airlock-type arrangement, and then you're in cat heaven. All kinds of cats doing all kinds of cat stuff, and boy do they love those lollies. Dave sat there with a huge smile on his huge face, an enormous tabby licking away. And I've never seen so many pretty Japanese girls in one place: some in groups, some alone, all in complete cat contentment.

It was the perfect way to relax before showtime. Summersonic, the festival we were playing, runs over two sites: one in Tokyo and one in Osaka and these were the only shows we were doing in the whole of Asia.

Arte was loving every minute, and so was I. We were

determined to. Who knew? We certainly wouldn't be back any time soon. If ever.

~

It was unbearably humid and unbelievably hot, even when we went on after midnight. We sweltered through the set, and it was all good, but just after we had started playing 'The Narcissist' in the encore, the vocal mics all went down simultaneously.

Normally when there's a problem, you explain to the crowd that there is a problem, and please bear with us, but it's hard to address a crowd without a microphone. Especially when it's a big one and when they are all making as much noise as they can.

This crowd, many of whom had travelled from far and wide to get a glimpse of the band, were at fever pitch, full momentum. With no singers to point their spots at, the lighting team began to illuminate the bouncing masses, which just made them bounce higher.

The guitar and bass seemed to be working perfectly OK, so Graham and I just kept on playing while the backline sound crew swarmed the stage in a panic: waving, pointing and scurrying.

Graham smiled at me and put a little flourish into the guitar part, and I threw it back at him from the bass and we carried on a-riffin' and a-jammin' just like we used to in the halls of residence at college where we met, just bouncing off each other. Like we were both nineteen again. Back then, it was just the two of us, and now we were plugged in to half a million quid's worth of twenty-first-century Japanese sound equipment and it was sounding crispy, and we just kept letting rip off the cuff and suddenly everything was working again and the vocals kicked in and the whole thing was just effortless and seamless and I was thinking about just how much I love Graham.

And the crowd merrily bounced themselves beyond exhaustion.

~

The guy who ran the big cool nightclub in Shibuya was at the show and invited us for drinks afterwards. It wasn't so much a nightclub as a neon-lit rainforest on a wraparound terrace in the sky. The city a carpet of lights, another galaxy floating way beneath us.

It was fun up there. Arte said it was even better than his birthday, at which point I told the hostess it was his birthday just to see what would happen, and then there was all kinds of flare twirling and cork popping, and a dance troupe giving it the full happy-birthday-to-Arte-works kicking off on the tables when my phone rang.

It was Geronimo.

He said the people who were going to supply the bread rolls for the Steakation Sandwiches in the Cheese Hub had changed their minds and did I know anyone who could supply five thousand bread rolls by Friday and I said I'd think about it.

By the time I woke up the next day, he'd already fixed it. Not only fixed it. He'd done one better. He'd persuaded The Oxford Bread Company to get involved. I was proud.

Oh, and it had stopped raining and the sundial was telling the time, he said.

~

Sunday was punctuated by phone calls from the ops team and the office, and I was fully braced for catastrophe every time – but everything seemed to be just about under control back in the parish by the time the band were bundled through the just-about-under-control fans gathered at the station, to board the bullet train to Osaka.

I've always loved Osaka and had been hymning its sci-fi scenery and spa delights to Arte.

'You're literally not going to believe it,' I sang. 'It's the most foreign place I've ever seen.'

But time was against us. We were whisked directly from the station mêlée and whooshed straight to a pristine palace of a hotel, all spick and span and silent and still. Same old swish as anywhere else.

Even when you're right there, bang in the teeming thick of it, Japan can feel out of reach. Osaka was definitely there, but it was as far beyond this bubble of suffocating opulence as the Feastival build, which was fully *en avalanche* on the other side of the planet.

The show was late, and the flight was early, so after we played I stayed up all night walking around deserted downtown with Arte, and felt Osaka waking up for the week. We took the first internal flight down to Haneda, and then all the way to Heathrow, and a car back to the farm just in time for a very late breakfast. It was Monday, and Feastival was open to campers from Thursday.

Over the Rainbow

~

Claire was in the Cheese Hub in high-vis, directing an army of teenagers equipped with mops, buckets and paintbrushes. All the kids' friends, basically. Some of them were clearly working hard and some of them were standing around on a dirty floor holding mops, looking confused. It's easy to spot the useful ones. Georgie was sixteen when she started, and now she was running the office.

Actually, Georgie had left Geronimo in charge of the actual office and set up a battle command HQ upstairs in the Cheese Hub. A big desk covered with laminated timetables split into fifteen-minute intervals. Checklists of duties, actions, deadlines – and it was all happening thick and fast. She was up to her neck in guest lists, VIPs, sponsors and dancers and had her own entourage of assistants and a queue of people lining up to ask her questions.

The twins were running a restaurant this year – they were going to be doing cheese on toast. I radioed them and found them in the tomato shed messing around with a conveyor oven and buckets of freshly grated cheese trying to perfect the recipe and feeding all their friends lunch.

It was all whirring like magic.

We agreed to switch to radio contact for internal communication. I gave Georgie my phone and my laptop and went back to the house to see what was happening there.

Flava Dave and the whole Cheese Hub catering dreamteam were in the Dutch Barn, which had transformed into a kitchen. They were chopping half a tonne of onions and hammering out half a tonne of minute steaks. There was a new kid going at it like mustard.

'Who's that guy?' I said.

'Yeah, that's Benedict, one of the twins' mates. He's useful.'

Time had run out on the expialidocious super-Frazzles.

I really thought I'd solved the problem with poppadoms – basically plate-size crisps. Just big enough for a Steakation Sandwich, a couple of Britpop bhajis and some of Claire's salad from the market garden.

Flava Dave and I had a slight difference of opinion there, though. He wasn't keen and he wasn't backing down. I'd been serving the kids their dinner on poppadoms and making them eat standing up, and it worked most of the time, but poppadoms are a little on the fragile side and there had been one or two catastrophic structural failures so we were back on the old bamboo.

I'd had a result with cheeseboards, though. I'd spotted some gargantuan crackers – family pizza-sized round Ryvitas – in Sweden, and I had managed to persuade Flava Diva to serve the cheese on those, like giant edible cheeseboards. We'd managed to get a pallet FedExed direct from Scandinavia. Claire's heritage tomatoes kept rolling off the sides and into the pickle or the butter pat, but it was progress.

~

We were all on track, but by Thursday it was a bit like sailing through a tornado.

I'd given Ruby and Violet the opening slot on the

main stage. There was keen interest from Sony Publishing and the head of the company was coming with a view to giving them a deal. So was Jim Abbis, who produced Adele's first album and the Arctic Monkeys' first album, and also Kasabian's. He was driving four hours to catch their first live performance, having heard demos of 'Superficial Love' and 'Echoes and Mirrors'. So there was a lot riding on it.

That afternoon, I found them in the Thresher Barn with Geronimo, who was keen to roadie for them as well as DJ and run the office.

We discussed the set list. They said they didn't want to play any of their old songs. None. None at all, they said. Just new ones.

Discussions became somewhat heated.

They're incredibly prolific, and apparently get bored with what they've done even quicker than Albarn does. Jesus Christ. I'd lost control of them already. They hadn't even made a decent recording yet.

They started rehearsing and began to draw a crowd of cheese munchers from the tomato shed and steak bashers from the Dutch Barn. Robert had arrived from Denmark and was looking on, captivated, along with the Wilson twins in a little professorial huddle in the corner.

It seemed to be working OK.

~

The house got fuller and time seemed to go quicker and we were all consumed – the whole extended family – in a hurdling hopscotch sprint to the finish line.

All Back to Mine

As the sun set on Thursday we were ready, we were willing and we were able.

We jumped on buggies for a grand tour of the site.

Miles of festoon lights, floodlights and spotlights illuminated every corner, and the entire farm was transformed into a still, silent and spectacular night garden.

12.

Totally Cherry

August

I woke at 6.00 a.m., five hours before doors, and drank my coffee outside. It was late August, high summer, and the whole site was ready to detonate into memory-making good times. The calm before the funstorm was strangely exhilarating.

I prefer not to know the weather outlook at this stage, but the national weather forecast was broadcasting live from the Cheese Hub on breakfast television so I couldn't really avoid it.

I could sense the presenter felt the thundercloud levels of electricity, too: the giddy anticipation of mountains of hard work coming to fruition. She was genuinely delighted to tell me we were in for a perfect summer bank holiday weekend of clear skies and brilliant sunshine.

~

I was still glowing in the warmth of her television smile as I posed for the photographers, spouted for the media and spoon-fed the ever-ravenous socials monster.

The old duck pond in Railway Field, which, I was

relieved to note, had never looked or smelled better, had been transformed into a scene from 'The Owl and the Pussycat'.

I was sitting in the bulrushes in a beautiful pea-green boat with an edible cheeseboard addressing the media when Beatrix arrived in a buggy with Saver Dave to take me up to the Cheese Hub for the bar staff briefing.

Are your wigs itching? I said. They were in matching blond perms. I had a lurid green bob.

The DJs Tall Paul and Seb Fontaine were hosting a syrup takeover in the Cheese Hub later in the day, so the dress code was: wigs.

I'd given Beatrix free rein on Cheese Hub merch. She's been running that shop since she was at junior school, and it's been going from strength to strength. She had been tie-dying T-shirts, ordering bubble guns and glitter cannons and busying herself with it for the entire summer holiday. She'd very sensibly invested half of last year's profits into a broad selection of party wigs, which were already proving popular before the gates had even opened.

Lucy, who runs all the bars in the Cheese Hub, had the health and safety briefing well underway when we arrived. The bar staff, about a hundred of them, listened on seriously in their tie-dyed T-shirts and disco wigs.

I'd come to talk to them about the secret drinks menu. I'd solved the problem of what to do with quarter of a tonne of honey and two tonnes of cherries: you turn them into a high-strength alcoholic drink.

~

Totally Cherry

Through James Graham – the chef behind the melty cheese doughnuts that had been a runaway success a few years back – I'd got in touch with a distiller, and I'd managed to persuade him to sell me cask-strength grain spirit. It's quite hard to get your hands on, that stuff.

Vodka is basically watered-down grain spirit. I didn't want to water it down, I told him. I wanted to cherry it up. He was happy to help.

Flava Dave and I got to work and made what amounted to a cherry and honey bouillon. We added it to the grain spirit and, six weeks of steeping later, it was beyond delicious and *très restaurant*.

I've never had anything like it, not even with Lemmy.

It tasted like cough medicine, but in a really nice way, and it was deceptively potent.

Once we'd done the tasting, we congratulated ourselves on how incredibly simple it had been to make – just three, well-balanced organic ingredients.

And then we had another glass, and we were marvelling to each other about how incredibly delicious it was, and marvelling that we literally couldn't taste the alcohol it contained, and then marvelling again about how something so incredibly delicious could be so incredibly easy to make.

Then we remembered it had taken fifteen years to grow the cherries and five years of bee disappointments to get the honey and we both suddenly found that incredibly hilarious and when we'd finally stopped laughing we moved on to the next item on the agenda, the pigs.

We went online and ordered a gigantic fridge for curing salami and had another couple of glasses of what we had christened 'Totally Cherry' and called it a day. A memorable day.

The next morning, I woke up wearing a miniskirt, and I couldn't remember why, but was still feeling remarkably jolly when a gigantic oven specifically for cooking sausages arrived, mid-morning.

I could have sworn it was a fridge we'd ordered. So could Dave. He was still quite jolly, too, but Sian the office manager couldn't see the funny side.

She was still trying to get the money refunded on the sausage oven, but Dave had by now successfully tested Totally Cherry's alcohol by volume and it was a little stronger than we'd expected.

~

I gave the bar staff the short version of the story because the gates had already opened. From command HQ, upstairs in the Cheese Hub, the whole extended family watched the crowds thronging in, full of excitement. The whole site sprang to life in the morning sunshine.

My first job was to welcome everyone, and introduce Ruby and Violet on the main stage.

I watched them from the sound tower with Jim the producer and the head honcho from Sony. They absolutely smashed it. But I couldn't stick around – I had to leave directly after the last song, the very second they finished, to meet the available kids and Flava Dave at the cooking stage, where we were attempting to demonstrate how to create the perfect family farmhouse lunch with accompanying cocktails for Matt Parker and his wife, Professor

Green, while talking about stars in general and certain specific consequences of special relativity – most notably time travel.

Claire was DJing in the Cheese Hub, and I wished her good luck over the radio comms. Georgie dropped me at the cooking-stage green room and was about to fetch Geronimo from the main stage – he was due to be mixing up all the cocktails – but he turned up before she left in a hot-wired buggy with Ruby and Violet.

'STOP STEALING THE BUGGIES,' I said, and added, 'That was brilliant, by the way.' But the twins knew they'd smashed it. They'd been chased across the arena by their newfound audience, done a whole bunch of selfies and been offered free lobster burgers by an ecstatic street food vendor, as well as free access to the bumper cars and a publishing deal.

~

By half past two in the afternoon I had done a Q&A, a DJ set, and cooked a seven-course lunch with cocktail pairings using all the best stuff from the orchards, the market garden, the polytunnel and the tomato shed. I was totally exhausted. I went back to command HQ to get some coffee. It was empty apart from a billionaire and an A-list Hollywood star. I said, 'Oh, hello,' gave them a bottle of Britpop, and lay down and shut my eyes.

For about three minutes.

It was Georgie.

'We've run out of cheese.'

'It's the Cheese Hub! We can't run out of cheese!'

'Yes, well, the cheeseboards are OK, but they shorted us on the mild cheddar delivery for the cheese on toast and it's been flying out the door.'

Flava Dave was dispatched to Bookers to get cheese. Benedict had risen rapidly through the ranks and was put fully in charge of Steakation Sandwiches while Dave was away.

I went down to check how Benedict was doing. He was doing just fine and I was only gone about ten minutes, but when I got back upstairs in the Cheese Hub, the place was absolutely packed with afternoon ravers: dignitaries from all walks of life, a headteacher, a newspaper editor, an earl, a duke, a whole bunch of film and television executives, mathematicians, models, musicians and chefs, all doing the Macarena and guzzling the Totally Cherries.

I left them to it and went with Matt Parker and his gang to the comedy stage to talk about cheese. The comedy tent was new and it was proving very popular.

I spoke a little bit about how cheese is made. I said that we normally make Blue Monday on Monday. Tuesleydale on Tuesday and Wednesleydale on Wednesday.

That sort of thing.

~

I caught up with Claire on the big wheel afterwards. The two-seater gondolas were literally the only place on the whole site where either of us could sit without being asked questions.

We sat there holding hands, enjoying not speaking.

Everything seemed to be going just fine. The whole

farm, the farm that we'd bought as a ruin on our honeymoon, the farm that we'd thrown our whole lives into, was full to bursting and full of joy.

We swung around in comfortable circles and agreeable silence for another go, and that was when we both realized we hadn't eaten anything all day. It was six o'clock and we were absolutely starving.

We went to get some cheese on toast from Arte and Gali as we hadn't seen them all day. They said we had to get to the back of the queue or everyone would think we were arseholes. The queue was massive.

In their development kitchen in the tomato shed, they had come up with a miraculous blend of three different cheeses that oozed and bubbled and flecked golden under the grill like nothing I'd ever seen. Now they were rolling it out.

The toast was perfectly crisp. The cheese was perfectly unctuous. They were dusting the finished slices with pepper then burying them with exotic salad leaves from the market garden and micro-herbs from the greenhouse, serving the finished item with a napkin and an ice-cold glass of apple juice.

It smelled incredible. It looked amazing. My mouth was watering.

The intense heat of the afternoon had mellowed to a pleasing, cool sunset and I seriously can't remember craving anything as much as I was now craving a slice of that perfect cheese on toast as the sun went down on a full day.

We put an order in, but before it arrived Claire was pulled off to meet and greet some VIPs, and I had to go in the opposite direction to meet a delegation from Iceland who were keen to take Feastival out to Reykjavík.

They were having a great time. They were raving about some incredible cheese on toast from right next door and insisting that I had to try it. I said I'd certainly do my best but there was still no sign of it when Saver Dave arrived with Georgie on the buggy to whizz me back to the main stage to introduce Sigrid, the headliner.

Sigrid was brilliant. I watched the whole set from the lighting tower then dashed backstage again to congratulate her and thank her. I said I'd see her in the Cheese Hub if she fancied it and bounced back there, full speed, on the buggy.

The syrup takeover was peaking. It was absolutely rammed with bouncing, flailing, rainbow-coloured hairstyles. I thought I had nothing left to give, but within five minutes I was going full 'Rasputin' with both grannies and Beatrix. Then Richie, Lydia and a whole bunch of the Blur crew showed up and joined in and on it went until curfew at midnight.

~

I had a lie-in until 8.00 a.m. Robert made me an omelette and Georgie brought the newspapers. We'd made a couple of the front pages.

Saturday is a slightly longer day than Friday. The event starts earlier and closes later.

So, at the risk of stating the obvious, it was another busy old day.

Problems, solutions, sponsors, artists, big wheel, guests,

'Rasputin': the whole family consumed but just about coping with the ebbs, the flows, the vast momentum of the circus.

~

Sunday was Sable's birthday and she was having a party in the Cheese Hub at three o'clock. One of the pastry chefs had made her the most wonderful cake anyone had ever seen. Standing there, she looked so pretty and so grown-up that I cried a little bit. Beatrix shut the merch stand and took the decks and got busy with the birthday bangers and the bubble guns.

Bea was a natural. In next to no time, the place was packed, and then Sable was up in the DJ booth with her, and they were both bouncing up and down and waving their arms in pure joy, abandon and release. Their brothers joined them and then Claire and I did too, then Bea called up the grannies and introduced them and let it rip with Chic's 'We Are Family'. It blew the roof off. We bounced and we bounced, and bar staff and ballerinas were dancing on the counters, the crowd was spilling way out, into the main arena.

This, I thought, without the shadow of a doubt, was the most wonderful moment in a year of wonderful moments.

~

After I'd done my final link from the main stage – thanking everyone for a wonderful weekend as the sun set and the sundial cast its final shadow – we cracked the nebuchadnezzar of Britpop that I'd been saving for a special occasion.

A nebuchadnezzar is a big bottle. Really big. Ideally, you want two footmen to tilt the liquid into your glass. The twins did a pretty good job. I figured I'd earned a drink and, my goodness, it was good with the cheese on toast.

I had to leave first thing on Tuesday morning. Blur were headlining a festival in Portugal – our final European show. Even as an item in the diary, it had looked like it would be a tricky corner to negotiate.

It was trickier than I thought.

~

It turned out that Britpop with just a little splash of Totally Cherry was the nicest thing anyone could ever remember tasting, and Sable's birthday party went on quite late.

I got to the hotel in Porto feeling utterly wretched. I opened my suitcase and it contained three belts and a pair of Claire's trousers and a small Evian bottle full of Totally Cherry. And nothing else.

Galileo was accompanying me on this jaunt. Everyone else had gone to bed for a week but he was keen to see The Prodigy, who were also playing.

'Blimey, look at this, Gals,' I said.

'Hmm. Mummy's trousers. Why'd you bring those?'

'I really don't know. Totally thought I'd aced the packing, Do you feel OK? Need a belt?'

'I'm fine, Dad.'

Totally Cherry

'Well, I can honestly say I have never, ever felt this bad. I'm pretty sure this is worse than after Mummy's fiftieth. Remember? When everyone was doing the splits?'

He laughed. I laughed as well and laughing made me feel even worse.

'We're gonna have to throw the book at this one, Gals. Call the spa and see what slots they've got. Grab every available one and book all the most expensive treatments. It's an emergency. Loofahs. Seaweed. Hot stones. Birch twigs. Whatever they've got.'

He managed to get us a 'hydro massage for couples', whatever that was, and there was some kind of mud-based pedicure and a deep-cleansing facial which we tossed for. I got the 'mudicure'.

'Food might be a good idea as well.'

The hotel restaurant had two Michelin stars but sweetbreads and parfait de foie gras weren't going to cut it. Not then.

'You know what this calls for, right?'

'Emergency Maccy's.'

'Reckon.'

Gali got busy with his phone, but by the time the rustling paper bags arrived I was already sound asleep. I slept right up until it was time to go to the venue, woke up, braising in perspiration and feeling even worse. If that was even possible.

'This is bad, Lyds,' I said, when I got to the dressing room. 'Feels like acute cheesehubitis again. It's a seasonal thing, I think. Never had it this bad, though.'

'Have you done a Covid test, love?' she said, reaching in to one of her many drawers.

I've never seen a Covid test so conclusive. It was like someone had drawn the second line on with a felt tip pen.

'Oh, bollocks.'

'Yeah. Maybe stay in here on your own. I'll put everyone else next door.'

I supposed it was almost inevitable when I thought about it. Apart from the odd moment I'd snatched on the big wheel with Claire, I'd spent the whole weekend being high-fived, kissed and embraced by friends, sponsors and strangers. I felt slightly better for the diagnosis, but I hadn't had the most restful weekend, and I could have done without additional complications.

The show had to go on, no question, and it did.

And I have to say, it was great.

Endless healing waves of euphoria from the crowd. There is no doubt that playing very loud music with dear friends does give you a lift, and, God bless them, they were all really sympathetic.

That was probably what made me feel best of all.

13.
Further Afield

September, October

Somehow, I got home. I took to my bed and stayed there for nearly two weeks. Fortunately, other than the Porto show, there was nothing in the diary for a fortnight. Literally, there was just a big cross and the word 'NOTHING'. I needed a chance to catch up with myself; to enjoy all the fruits and all the vegetables of the summer harvest.

I should have been feeling on top of the world, but I was ragged. Flattened for what felt like eternity.

By the time I was back up on my feet, summer was well and truly over.

I had more time to contemplate the future now – but what the future held for Blur, beyond this last scheduled run of shows, was impossible to say. The kids were asking me, and I couldn't tell them because I just didn't know. We'd all be going back to our own all-consuming separate worlds, if the past was anything to go by.

I supposed the future of the band hung to a large extent on how we were all getting on when the music stopped at the last concert, and also on how the record performed.

We all knew we'd made a good record, but you never

really know until five or six years later how good it was. That's how long it takes. It's very hard to tell which songs will endure. Everyone in the band had thought 'Song 2' was a B-side but it was our first proper hit in America. And it was 'Ghost Ship' from the previous album that had risen to become one of the band's most streamed songs, and that hadn't even been a single.

Damon is right: if you can't keep doing stuff that's up there with the best you've ever done, or better, that's when you have to stop. Or you just become your own tribute band. And no one wants that. Farms, festivals, bands – whatever you're doing: if stuff ain't growing, it's time to call it a day.

But that was the future. Right now, we had a grand finale on our hands and the grandest of finales it was shaping up to be. As autumn turned to winter, Blur were booked to play a final run of festivals across South America. Prior to that, by happy coincidence, Feastival was launching under licence in Peru.

~

I absolutely love South America. Absolutely all of it, and absolutely everything about it. From top to bottom, coast to coast, it's a place of endless wonder, endless enchantments, and endless possibilities. It's another world, altogether.

South American audiences throw themselves into Blur's music with pure, open-hearted abandon, quite unlike anywhere else, and it is with that same sense of unbridled enthusiasm that I embrace the entire subcontinent.

Further Afield

It's partly down to the food, too: it's mind blowing. South America is where chocolate comes from, and tomatoes, and potatoes and cigars, obviously – but that is the tiniest tip of a vast iceberg of undiscovered delicacies. You could spend a lifetime on the food cultures of Peru alone and still not discover everything.

Peru has the Pacific Ocean on one side and the Andes and the Amazon on the other. It's got lost tribes and it's got just about every microclimate going. I'd been there a few times by now but my initial impression of Lima, the capital, was lasting: this was a place of immense scale and intense colour.

It was like the divine hand had turned the contrast, temperature and humidity controls right up to maximum at some point during the last Ice Age and they'd been stuck like that ever since: all brilliant blues forever and brilliant greens running wild in a mad outdoor orangery.

From previous travels in South America, I'd got to know a Peruvian guy, Guillermo. Got to know him very well, in fact, and liked him immensely. I'd put my life in his hands on numerous occasions.

Guillermo's what's known as a fixer: someone who seems to know absolutely everyone and knows how to get just about anything done. There is a sense of wizardry about the man.

The first time Blur played in Lima he'd taken great delight in showing me around. We really hit it hard: from grand restaurants to everyday cafes to street food vendors, all exquisite and all engineered cleverly by Guillermo so that not just every dish but every last ingredient on every last menu was entirely new to me. It was largely thanks to him that I'd realized Lima was the perfect place to take Feastival.

Sadly, Guillermo was travelling and wouldn't be there when we arrived, but the plan was to decamp the entire office to Lima for a week or two, give the Peruvians the full Cheese Hub experience – DJs, cheese on toast, professors, the whole circus – and then fly on direct to Mexico City to meet the band and kick off the Blur tour.

~

On my first day back in the office, everyone was raring to go.

I started wading through the six million emails that had accumulated since I'd last looked and was relieved to see I hadn't missed much from Peru. Nothing at all in fact.

As soon as I came up for air I shouted at Georgie's office door, 'Blimey. Thanks for handling Lima, Georgie. You're marvellous.'

She appeared, looking confused. 'I thought you were handling that?'

'Literally all I've done is watch six seasons of *Breaking Bad*. That is my news in its entirety.'

I emailed the Peruvian promoter, but I didn't get a response.

Not until the next day, when I heard from someone else, who called me to say he wasn't a hundred per cent sure, but it sounded like the Peruvian promoter was either maybe in court or possibly in prison and it was probably best to pull the plug at this stage.

It was a real shame. A lot of diligence and hard work, all come to nothing. And I was still shaking off coronavirus's final throes of doom and depression. And Ruby and Violet were getting frustrated with me, too.

They'd been in the studio with Jim, the producer who'd come to Feastival to see them. They'd also tried working with one or two other producers on Sony's roster. The girls could do all the difficult bits – be effortlessly cool, nail a top line with a pretty tune and words that made you cry – without even thinking about it. But they were struggling to land on a sound they were happy with.

They mainly wrote to a piano, and a piano accompaniment is fine for a cocktail bar but not for a massive global pop record. They really just needed time to develop and find themselves a bit in the studio, but it seemed I was now the cause of all their frustrations rather than the answer to all their prayers.

I felt gloomy. It was raining. I'd lost the previous fortnight and now the coming fortnight had gone up in smoke.

~

Claire cheered me up. During the time I had been flat out with *Breaking Bad*, she told me, she'd built a stumpery.

I didn't know what that was.

It was a big cocoon of fern- and toadstool-sprouting tree stumps with a big bonfire and a small kitchen in the middle.

Over the Rainbow

It was a nice place to watch the sunset and linger under blankets and stars sipping hot chocolate.

Claire suggested we go to Somerset for a couple of days of recuperation.

When we got back, a couple of weeks later, I was completely restored, and it was time to go to Mexico.

14.
Finale

November

I was asleep before wheels up, and when I woke, I was in the biggest and most exciting city on the planet and it was Friday and it felt like springtime and it was nearly Christmas.

There were a lot of Blur fans at the airport and a lot more at the hotel, the Four Seasons. I had three local security guards, plus Saver Dave, and Claire had two as well, which meant we needed three cars to get to the hotel.

Claire went straight to bed and ordered room service. Mexico City is ten thousand feet above sea level. The altitude combined with the jet lag can hit quite hard – but I was wide awake.

Saver Dave called and said the lobby downstairs was rammed with fans, and the record company and publishers were wondering if I wanted to go out for a mezcal or two at the place I'd liked last time.

I said I'd love to, but had booked a massage in the spa. Then I switched my phone off, threw it on the bed and snuck out on my own via the service lift. An unscheduled solo 117.

Over the Rainbow

I walked in a straight line, soaking up the city's warmth, until I got to the nearest place I liked the look of and I went there and had the suckling pig.

Then I went to the barber's next door and had a hot-towel cut-throat-razor shave. It was a bit like a gentlemen's club. They were all smoking cigars and drinking bourbon from a gimballed jeroboam. I wasn't in a rush to go anywhere, so I stayed for a while.

It's not like the Four Seasons doesn't do a good dinner and a good shave, but it's important to get out and about. And to meet new people as well, which can be hard when you're in a strange city and surrounded by armed security guards.

~

It had been a few weeks since the last Blur show, so we had a soundcheck and a rehearsal at the crack of dawn, just before the arena opened. This was the biggest arena yet. Possibly the biggest one I've ever seen.

And the biggest problem with sound-checking just before doors at a festival is that that's when the slurry wagons come to empty all the Portaloos.

The PA was incredibly loud and the morning sun was incredibly hot and the heat was amplifying the stench from the slurry slurping, which was happening at a frenzied pace and on a biblical scale.

Richie rallied round with shades, sun block, ear plugs and also some nose plugs, which he'd cleverly fashioned from cigarette filters.

It could have gone either way. But, somehow, playing

Finale

really loud music was the perfect balm for the jet lag, the altitude and the after-effects of a four-hour shave. It was practically transcendental. When we'd run through everything, note-perfect, I floated back to the dressing room in a state near to bliss.

But it was busy in the dressing room. It was full of management and record-company and PR people. There were a lot of media requests and an invitation for lunch with the ambassador at the embassy.

I spoke to the cameras for the evening news but swerved lunch with the ambassador and went back to the Four Seasons as Claire had booked us the Mexican signature couple's massage. There was a whole team of masseuses waiting. As some lit small fires and wafted smoke around in the dark, and others banged drums, all of them chanting, another explained that the Mexican signature massage was all centred around tequila and chocolate.

Footladies appeared bearing candles and ceremonial tequila shots. I waved mine away saying I had to go to work shortly. They poured tequila all over us and vigorously rubbed it into our skin and then they slavered us in melted chocolate and then I must have fallen asleep, and when I woke up, I felt magnificent.

~

Claire wanted to see Jungle, who were also on the bill, and was worried we were going to be late to site, but I reassured her it would all be fine as we were leaving the hotel hours before their set started, and besides, we had a police escort.

Over the Rainbow

We met security at the lift lobby just before nine, as planned, took the service elevator to the car park, and waited for ten minutes for everyone else to turn up.

The blacked-out vans were a bit big for the car park and had to do ten-point turns to get out, and there were so many of them involved by this stage, all pirouetting like mating crabs, they somehow created their own gridlock within the car park itself.

Fifteen minutes or so later, we had joined the police escort at the car-park entrance, and the entire convoy raced down the block, at which point we joined the main city gridlock.

The biggest city in the world can get quite busy on a Friday night, especially at Christmas, and especially if there's a big event on, so we all sat there, sirens wailing, lights flashing, as outriders perilously rode red lights and attempted to stop oncoming traffic by flailing arms around.

We finally got to the dressing-room compound at midnight. Claire disappeared into the arena to meet some friends and watch Jungle, and I was just settling down to a blissful cup of Lydia's finest when the ambassador, who was clearly most keen to catch up, arrived with his entourage of diplomats, spies and pretty women.

There was a very fit young Cambridge graduate who'd just got off the plane from Israel and seemed to have all kinds of questions that needed answering and by the time they'd all been ushered out it was nearly two in the morning, and fifteen minutes to showtime.

Finale

It was good to be back up there. The massed denizens of that fair city were singing every word and crying their eyes out and throwing their beer: a good day in the office, all told, but I wasn't getting much from Albarn on stage, couldn't catch his eye at all, in fact, and couldn't work out why he was ignoring me. I kicked him up the bottom to see what would happen, and still nothing. He still didn't look at me.

I didn't know why, because you never really know exactly what it is about you that the people you know best really love and really hate about you. It was painful.

Damon Albarn is a truly exceptional human being. I'd come to realize that more than ever over the past few months. Supremely gifted musically, and with the work ethic of all true leaders. He'd given absolutely everything he had to give to this album, and this tour: gone all-out, win or bust.

I wondered, then, if he'd had enough, and that was it for another ten years, or maybe forever. I turned my head to the bouncing crowd, arms aloft ecstatic as far back as the floodlit horizon, and I felt sad.

The final cymbal crashed on the final cadence, which I embellished with a particularly satisfying flourish up and down the harmonic series on the A string. When I looked up, the singer was nowhere to be seen. He'd literally pelted off the stage like he couldn't wait to go.

He wasn't in the dressing room, either.

The crowd was going crazy. It was deafening. I've rarely felt so deflated.

'Where's Damo, Lyds? What the hell's up?'

'Food poisoning! Miracle he got through the full two hours. He did a 118. Got a medic with him, escort, should be fine.'

Her radio crackled and confirmed that he was, indeed, just about OK, but it was a minor miracle that he'd fronted an entire show and brought the house down.

What a man. What a machine. What a relief.

As we waited in the cavalcade for the drummer's girlfriend's guests to be located so that we could fire up the sirens and the strobes, I reflected that everyone had dug deep. I thought about Dave being pushed to the stage in a wheelchair at Wembley, and then nailing the best gig we've ever done. I thought about playing in Porto with Covid. I thought about Graham saving the power outage in Japan. It had been such a great source of contentment just being with Graham again. He makes me feel good, makes me laugh, always has done.

The band getting back together had brought many gifts, but the most precious of these had been enjoying all of their company.

They're my oldest friends. The cunts.

~

It was Sunday. I got up early because the Four Seasons breakfast buffet is a thing of great wonder, and if you get there as it opens, especially on a Sunday, it's all pristine and wonderfully silent, like it's all been laid out just for you.

Claire said in no uncertain terms that she didn't want breakfast so I ambled down on my own.

Finale

Smog, the band's head of security, was in the lift when it arrived on my floor, also keen to make the most of today's smorgasbord.

'Are you going to the pyramids or back to the festival?' he said.

Mexico's got the best pyramids, but The Cure were closing the festival. The first proper gig I went to was The Cure at Poole Arts Centre. The Breeders and Noel Gallagher – who I was keen to book for Feastival – were also on the bill. It was a stronger line-up than Glastonbury.

The lift doors opened onto a jungle glade – a roofless atrium of deep rainforest, all birdsong, butterflies and bananas – bang in the middle of downtown. As the omelette chef busied herself, Smog mentioned that The Cure were having a party at the hotel and the bar would be re-opening at 4.00 a.m. when they got back, and everyone was going, and they'd love us to join them.

~

In the end, Claire and I did nothing all day and went for a walk in the evening. Doing nothing is almost impossible on tour and even going for a walk is a palaver.

Word had got out that I'd slipped my reins the first night and I'd been politely asked not to do it again, so we took the full security detail. They were pretty good, all huge, all ex-special forces, Dave said. The vehicles followed us, too, all three of them, hazards flashing. On foot, a navigator and a tactical road crossing specialist led the way, plus two front and two rearguard, with Dave walking alongside Claire and me.

It was a strange echo of the two of us marching all the kids around London, back when they were little. Somehow, we were the little ones now.

There were a million things just waiting to happen. A thousand different nights about to unfold. From them, we selected ant egg caviar, grasshopper guacamole and bed. Because the following day was a big one.

~

We were flying into Bogotá to do a show on my birthday. We needed security that day. Crazy scenes. Fans besieging the hotel as we left, fans outside the airport, inside the airport, airside. Even at the pointy end of the aeroplane.

I had my own bass fan in the row behind.

'You! You the reason I make bass play!' he told me with infectious delight.

I fell asleep upright, smiling, while the plane was still at the gate and woke up, still upright, still smiling, in Bogotá some hours later.

The last time I'd been to the Colombian capital I'd had a sizeable film crew in tow. We'd been met at the airport that time by armed security: fingers on triggers, eyes darting, and I'd been escorted in a fully bombproofed vehicle directly to the heart of government to be given a calmly administered but very thorough roasting by the boss regarding a flippant comment I'd made about how much cocaine I'd taken in the 1990s.

After I'd stopped taking it, which was just before I met Claire, I'd said how pleased I was that it was all behind me

Finale

now, as it most likely would have killed me because I had, quite frankly, taken it quite a lot.

So much, in fact, that the president had been compelled to write to me personally, inviting me to come and see what a mess the drug was making of the country he was running and exactly what he was up against, hence the documentary film crew.

I'd really liked him, and his vice president, and I'd found the country the most beautiful place on earth.

I'd apologized and done my best to make amends, and was relieved to see that I'd been stood down from bombproof to a standard bulletproof showbusiness-level security detail.

I woke up at dawn. I felt great. It was summer in paradise, and it was Christmas time as well, and it was also my birthday so Claire had to be nice to me all day.

But there was a lot to deal with. We'd missed Monday, because we'd been flying all day, and even though it was dawn in Bogotá, it was already lunchtime back in the English winter. I spent an hour on the phone to the farm office, then Claire and I had to Zoom with the twins as they were due to submit their college applications. And on it went.

I was getting a bit restless when my phone burbled: a message from my Peruvian friend Guillermo, saying he'd just got into town from Argentina, where he'd been to meet the new president, and he'd seen all the Blur posters.

I called him right back. He said he just had to go for a quick meeting with the Colombian president but he'd love to come to the show and we should definitely go out later.

'I tell you, man, there's this new place. You gonna love it. Issa fucking crazy place, man. I gonna see you later.'

And then management phoned, and said they thought it would be really nice to have a combined end of tour party and birthday party all rolled into one, and the promoter had found us a venue, and it was all sorted, and everyone was coming.

~

And it was, indeed, a good night. Saver Dave confirmed that on the way to the airport the next day, shaking his head and smiling as we pieced it all together.

Guillermo had rocked up with the head of JUNGLA, the Colombian counter-narcotics special force, and we'd got stuck into a bottle of Amazonian hooch that Damon had given me.

The party venue that the promoter had booked was, by chance, the very place Guillermo had mentioned earlier, and it was hard to imagine a sweeter spot. Salsa band. Cocktails. People of all ages dancing without a care in the world, like there was no tomorrow, and it had been wonderful to catch up and kick back with the crew.

'And I didn't take my trousers off, right, Dave?'

'No, that was Matty took your trousers off.'

'Not my bad?'

'No, they were all trying to give you the bumps and it just ended up with a big heap of people on the floor.'

'And I got to play "Blue Moon"! On the piano! At the hotel!'

Finale

I like to sing 'Blue Moon' on my birthday. My dad taught me how to play it.

'Yes, you did. Hotel weren't too happy about it, though. It was quite late.'

'Well fuck 'em. It was my birthday.'

~

Getting all the way to Chile was never going to be that much fun following a night like that, although we were flying under diplomatic cover. But that just means you queue up with diplomats rather than queue up with everyone else. Once we were past security, we all stood like cucumbers in a crate on a bus for half an hour in sweltering heat.

I have felt better than I did when we finally boarded the plane.

It was absolutely packed and we mucked about on the apron for another two hours. It was so hot that the plane was too heavy to take off on the short runway, so they had to take some bags off. All in all, I'd had better mornings.

But then, finally, we were flying again.

From feeling completely trapped I suddenly felt completely free, balanced on air and hurtling over strange horizons at breakneck speed. Over ocean, mountains and endless jungles, where the sunlight picked out quarries here and there and lit them gold. On and on, into the darkness, jewels of countless unknown cities floating past, way underneath.

~

Over the Rainbow

Santiago, the Chilean capital, had grown beyond recognition since the last time I had been there. The hotel, along with everything else I could see, was brand spanking new: an inviting maze of super-modern skyscrapers sprouting out of the green canopy. One aspect of the suite's forested terrace gave on to a complex of pools and fountains, another on to a wedding, perhaps a billionaire's daughter's, with twirling burlesque fire jugglers, lilting beats and pan pipes.

It took a while for the luggage to arrive. Happily, our bags hadn't been removed from the overweight aircraft, but I had received quite a lot of cheese for my birthday and the hours spent on the tarmac in Colombia's scorching heat had rendered it beyond diplomatic immunity by the time it landed in Santiago.

Saver Dave said they had been quite cross in customs, particularly about the one that resembled coiled rope, and was as large as a tractor tyre. They'd been keen to hang on to that one, which was a bit of a blow, but not enough to ruin the balmy, brightly lit evening beckoning with open arms.

~

Each day felt more precious as the Blur Doomsday Clock ticked down and down, faster and faster, until the final show, in Argentina.

The crowd sang so loud that Graham couldn't hear his guitar. The Argentinian economy was toast, but they'd just won the football and there was a brand new guy in charge. There was the promise of a better future

Finale

and music was the prism that all that promise flowed through.

What a way to finish.

~

The dressing room was rammed and quickly became a party. We all decamped to another venue to join another, larger, party and by the time Claire and I left, haring back to the hotel through the deserted streets under full escort at terrifying speed, there wasn't much time for sleeping.

Dave woke us up. It was time to go. The final 117.

I felt horrendous. It had been a big old week.

'Jesus, Dave.'

'Yeah. I've felt better, I must say.'

I stumbled out of bed. Pulled a Coke and a Fanta out of the minibar and offered him the choice. He smiled and took the Coke.

I necked the Fanta in pretty much one swallow, pulled the curtains wide, and stumbled onto the balcony in my pants. The sugar started to work its magic. I turned to face the brilliant sun and leaned my head back, closed my eyes and opened my arms wide. As the heat warmed my bones I became dimly aware of a kerfuffle far below.

I looked down and there were hundreds of people waving, cheering and chanting my name. All of a sudden, I felt much better.

And then it was all over.

15.
Next Year

December

Except, nothing's ever really over. It's just the start of the next thing.

Claire and I arrived home the day before the kids broke up for Christmas to another full-on shituation. Somehow, while we were away the cats had all apparently come under the impression that the kitchen, the whole main kitchen, the very heart and soul of the farm, was a big old cat toilet.

Claire's mum was dealing with it, valiantly: Marigolds on, trying to rattle a cooked turd out of the toaster I'd got everyone for Christmas as we walked in.

I told her I'd had an anonymous birthday card that read simply, 'Home is where the cat shreds the sofa and the dog shits in the kitchen but love prevails.'

She laughed, and said, 'You want a cup of tea, love?'

'I'll make 'em.'

Claire had disappeared into the garden. I found her in the tomato shed. She'd put gloves and an apron on and was busy at the propagator.

'The digitalis need potting on,' she said. 'What the hell have they all been doing?'

Mist gathered at the bottom of the valley and evening fell. By the time we'd dealt with the digitalis the stars were out. We walked back to the house and sat for just a moment on the bench outside the kitchen.

It was cold and silent and still.

~

The next day, I sent Damon the clips of Ruby and Violet singing 'Echoes and Mirrors'.

'Yeah,' he replied, 'bring them down to the studio.'

Acknowledgements

With thanks to my family, to my brothers and everyone in the office and on the farm, Penguin Books, Eleven Management, IMG and the many, many brilliant organizers, chefs, musicians and creatives who work so hard together each year to make Feastival happen.